A Naturalist's Guide to

Butterflies

of Britain & Northern Europe

TED BENTON

JOHN BEAUFOY PUBLISHING

First published in the United Kingdom in 2017 by John Beaufoy Publishing Ltd
11 Blenheim Court, 316 Woodstock Road, Oxford OX2 7NS, England
www.johnbeaufoy.com

10 9 8 7 6 5 4 3 2 1

Photo Credits
Front cover: Swallowtail. **Back cover:** Lapland Fritillary. **Title page:** Adonis Blue. **Contents page**: Peacock.
Main descriptions: photos are denoted by a page number followed by t (top) or b (bottom). All images were
taken by Prof. Ted Benton with the exception of the following:
Prof. Alan Dawson 107b. **Dr Bernard Watts** 73b, 79b, 91b, 105b, 107t.

ISBN 978-1-909612-45-7

Edited by Krystyna Mayer
Designed by Gulmohur Press, New Delhi

Printed and bound in Malaysia by Times Offset (M) Sdn. Bhd.

·Contents·

INTRODUCTION

Accounts of the butterflies of Europe differ in their estimates of the number of species that inhabit the continent. An estimate of butterfly numbers in this region is partly dependent on:

- Where the geographical (or political) boundaries are drawn.
- Whether scarce vagrant species are included.
- The decisions an author makes regarding disputes over whether some closely related forms should be treated as separate species, or 'lumped' together as a single complex species.

Despite this, an estimate of somewhere between 400 and 450 species can be considered to be reasonably accurate.

ABOUT THIS BOOK

This book includes descriptions and photographs of just 158 species that occur in northern Europe. They include all those that are resident in, or regular migrants to, the following countries: Britain, Ireland, Belgium, Luxemburg, the Netherlands, Denmark, Norway, Sweden, Finland, Estonia, Lithuania and Latvia. The book also includes almost all species that are likely to be seen in northern France, Germany and Poland, and the great majority of those to be found in Europe to the north of the Alps. There is just one proviso about this. Butterfly distributions are, like most of nature, always dynamic. Some species are in decline and contracting their geographical range, while others are colonizing new territories. By the time you read this, some species may have become extinct in particular countries, while others may have extended their range. Under the influence of climate change, a number of species have been extending their range northwards – for example from the Baltic states into southern Finland – while a few northern species seem to have abandoned some localities in the southern part of their range.

The accounts of the butterflies covered here include the European distribution of each species, the appearance of both uppersides and undersides of the wings and, where necessary, differences between males and females. There are also brief indications of the habitats each species favours, often characteristic aspects of behaviour and, where known, the plants that are used for food by the caterpillars. The accounts of each species are organized according to the family groupings to which they are currently assigned, and the order in which the families are presented is a compromise between up-to-date evolutionary thinking, and the order given in other books on butterflies that the reader might wish to consult.

Each family has a short general introduction, followed by accounts of the northern species in that family. In the case of one very large family (the Nymphalidae), it was thought convenient to divide the treatment among two distinctive subfamilies (the Nymphalinae and Satyrinae), which were given the status of families in many earlier books. Each species is referred to by two names, its scientific name, and a name commonly given to it in the English language. Non-English speaking countries have their own vernacular names for the butterflies, but the scientific names allow clear identifications across language barriers. For this reason it is desirable to overcome any tendency to resist using the scientific names: they are much easier to learn than many people assume.

The first word in the two-part scientific name tells you the genus to which a butterfly belongs: that is, what group of closely related, and usually similar, species it belongs to. Compare, for example, the skippers in the genus *Pyrgus*, or the fritillaries in the genus *Boloria*. The second word in the scientific name tells you which of the members of that group this is. In this book, the order of species accounts within each genus follows the alphabetical order of the first letter of the species name.

All the photographs were taken of free-flying butterflies in nature. This means that some species are, for example, represented by shots of their underside only. This is because they only settle with their wings closed – although very occasionally it is possible to take a clear photograph of such butterflies just taking off, or even in flight, showing their upperside. Where males and females are strikingly different from each other, where both undersides and uppersides are needed for identification, or where it is desirable to show some behavioural trait, several images are provided of the same species. Unfortunately, in a book like this it is not possible to show the early developmental stages of the butterflies, despite their fascination.

LIVING WITH BUTTERFLIES

Butterflies are among the most widely admired of all living creatures, celebrated in literature and the visual arts, valued as residents in urban parks and gardens and subjects of endless fascination for those who study them. Until evidence of serious decline raised public alarm, in the latter part of the last century, it was quite common for people to create collections of pinned butterflies. In the 19th century there was a thriving market in rare butterfly specimens to feed the demands of specialist collectors, and in some parts of the world this persists even now. However, for most of us today, the emphasis is on protecting the butterflies and their habitats. Many people manage their gardens to attract butterflies and other insects, and in the UK, for example, thousands participate in regular walks to record butterfly numbers so that their conservation status can be monitored from year to year.

BUTTERFLY ANATOMY

Just about everyone can recognize butterflies when they see them, but what are butterflies? They are, of course, insects, sharing with other insects a body comprising head, thorax and abdomen. The head carries important sensory organs – the large, compound eyes, smaller ocelli and palpi (paired pads below the eyes that are sensitive to chemical scents and flavours), and a pair of antennae, which are important organs of chemical sense as well as touch. The head also carries the furled double tube that forms the tongue, or 'proboscis'. This is used by a butterfly to suck up fluids – mostly nectar, but also ant honeydew, water and other fluids. There are three pairs of legs attached to the thorax. Like many other insects, but certainly not all, butterflies have wings, which are also attached to the thorax. There are four wings, and these are coated with minute scales, some of which are pigmented, while others are so structured as to break up and refract incidental light to produce various hues.

The vast order of insects to which the butterflies belong is the Lepidoptera (scaly winged), but the butterflies make up only a small part of the order – the moths also belong to it and are far more numerous. Conventionally, the butterflies, with six families represented in northern Europe, are distinguished by:

- Their habit of flying in the daytime.
- The clubbed tips to their antennae.
- Their habit of roosting with wings folded up over the body.
- The lack of a connection between the forewings and hindwings.

There are, of course, exceptions – for example moths such as burnets, which are brightly coloured, have widened tips to the antennae and fly in the daytime.

LIFE CYCLES OF BUTTERFLIES

While it is the adult butterflies that catch our eyes and charm our walks in the countryside, it is important to remember that the butterfly is just one phase in the life of the insect. Butterflies share with flies, bees, beetles and some other insect groups what is called a 'complete metamorphosis'. This means that the immature stages in the lives of butterflies are each very different from each other and from the adult butterfly. The early stages in the lives of grasshoppers and crickets, dragonflies and some other groups show many of the external features of the adults, and grow by shedding their skin several times, each time becoming more like the adult.

THE EGGS

Butterflies and moths begin life as eggs. These may be laid singly or in clusters on plant leaves or buds, or in some cases may just be scattered over tufts of grasses.

The microscopic structure of a butterfly's eggshell is astonishingly complex and often beautiful, the shape being very different across the various family groups. Inside the egg is a tiny embryo that first develops using the internal food supply until it is ready to hatch as a minute larva, or caterpillar.

The egg of an Orange-tip butterfly.

A Scarce Fritillary laying its cluster of eggs.

THE CATERPILLARS

A caterpillar is elongated, tubular or slug shaped, made up of a series of segments with a head at the front. The head has simple eyes, short antennae and mouthparts adapted for chewing plant tissue. The front segments of the body bear three pairs of legs, and several of the rear ones have soft 'pro-legs', with another pair at the rear end. These allow the caterpillar to grip the plant surface where it rests or feeds. Butterfly caterpillars have a wide variety of colour patterns, most of them evolved to give camouflage, so that greens and browns are the most widespread, often with irregular patterns that break up the appearance of the caterpillar's shape. In other cases (for example in many of the Nymphalinae), there are strong spines or bristles projecting from the body, and in yet others, bright 'warning' colours that deter would-be predators. The caterpillars are often distasteful or toxic, or maybe just mimic other insects that are. To make matters more complicated, some caterpillars change shape and colour pattern as they develop. For example, the young caterpillar of the Swallowtail resembles a small bird dropping, but later it develops a bright pattern of green with black-and-orange bands and is able to extrude an offensive-smelling organ when threatened by a predator. Unless it is under attack, the caterpillar's main business is to feed and grow. It grows by successively shedding its skin, becoming larger each time.

Caterpillar of the Orange-tip.

Caterpillar of the Silver-washed Fritillary.

Early-stage caterpillar of the Swallowtail.

Full-grown caterpillar of the Swallowtail.

THE PUPAE

When fully grown a caterpillar becomes less active, and attaches itself by a silk pad to some firm structure (or in some species, spins a cocoon). It sheds its skin for the last time as a caterpillar, making a transition to a new phase in its existence. This time the body is wholly encased in a hardened 'shell', leaving only tiny holes along the side for the exchange of gases. This new form is the pupa or chrysalis.

Outlined on the surface of the pupa are the shapes of the head, thorax and abdomen of the future butterfly, together with legs and antennae aligned together below, and small stubs representing the wings. This is sometimes referred to as a 'resting' stage, because the pupa has no means of locomotion (though it can usually 'wiggle' its rear end if provoked). However, what is going on inside is an astonishing transformation. The cellular structure of the caterpillar is 'dissolved', except for several groups of cells that multiply and differentiate to form the entirely new anatomy of an adult butterfly. This process can be

monitored from the outside, as towards the end the eyes show through the outer skin of the pupa, as do the colours of the wings and the other external features. Eventually, the pupa is broken open at the head end, and the new butterfly works its way out into the world, hangs from a suitable perch and pumps blood through the veins in its stubby wings. In this way the wings expand to make the shape of the adult butterfly wings, and become dry and firm. Now the insect is ready to take its first flight.

Pupa of the Meadow Brown.

BUTTERFLY BEHAVIOUR

In common language butterfly behaviour is the epitome of aimless flitting from one thing to another, though Mohammed Ali used a more complimentary analogy: 'float like a butterfly'. In fact, like all living things, butterflies have to face steep challenges in pursuit of their twin aims of survival and reproduction. Little of their apparently random behaviour really is random.

FEEDING

Once butterflies have emerged from their pupae and dried their wings, they seek out food. For most of them this consists of nectar, which they suck up by extending their curled-up proboscis and probing the inner structures of flowers. Floral structures are highly variable, often showing modifications to attract and reward insect pollinators. The nectar they secrete is the main reward for butterflies, while the flowers visited may gain from the pollen incidentally brushed onto the butterfly's body being transferred to the female part of the next flower visited. As especially long-tongued insects, butterflies can access the hidden nectar in flowers with long, fused corolla-tubes, but they often also visit flowers of plants such as brambles, which are shallow but have copious supplies of nectar, or knapweeds and thistles, which also have shallow florets, but have many clustered together, so reducing the energy costs of flying from flower to flower. Butterflies use their compound eyes to locate the colourful blossoms, but at close quarters their acute chemical sense organs, located in the antennae, palpi and feet, do the fine-tuning.

Large Skipper showing long tongue.

REPRODUCTION

The second main priority of butterflies is reproduction. In butterflies as in most animals, it is the males that spend much of their time in search of potential mates. They employ two main strategies in this respect. One is to patrol a regular route around features of the habitat, such as shrubs and patches of flowers where females might be located. A familiar garden butterfly, the Holly Blue, uses this method – often persisting for several days before the first females emerge. Other butterflies use an alternative strategy – they find a prominent perch and lie in wait. Butterflies that use this method are also usually territorial, so fly up to intercept other males (as well as butterflies of other species), engage in a spiral dance, then return to their perch. Some male Speckled Wood butterflies occupy sun-spots in woodland for this purpose, but can also revert to patrolling according to circumstances.

When a male manages to find a potentially receptive female (usually one that has recently emerged) there is often a courtship flight during which he releases pheromones from the areas of 'scent scales' present in specialized areas on his wings. In some skipper butterflies these are located in folds at the leading edges of the forewings, while in many of the browns (Satyrinae), they are located in broad bands across the forewings. If successful, the male settles next to the female and curves his abdomen around to make genital contact, and the pair mate back to back, wings closed, often deep in vegetation or, in the case of woodland butterflies, high in the canopy.

The importance of chemical communication is increasingly recognized, with 'antiaphrodisiac' pheromones being placed on females of some insects by males to reduce the likelihood of their mating again before eggs fertilized by them are laid. In some butterflies, notably apollo species, the males apply a physical barrier to the

Mating pair of Marsh Fritillaries.

female abdomen to prevent further matings. Instances of females rejecting male advances are quite often seen. The male hovers over the female, while she opens her wings and raises her abdomen with her rear aperture open. This looks like an invitation, but it is not. However, amorous males often take quite a time to get the message.

After mating, it commonly takes some days for the eggs to mature. Then the female's priority is to locate suitable places to lay her eggs. In most species this involves

A female Orange-tip rejecting an amorous male.

finding the preferred host plants of the caterpillars, and in some specialized species it involves active searching – female butterflies can often be seen 'treading' the leaves of plants, identifying them through touch and taste. Even when the correct plant is located, females are often very choosy: the plant must be in the right position (often exposed to the sun, but not likely to dry out), and in the right condition (fresh young shoots are frequently chosen, or in other cases, flower buds). In some species the eggs are not usually laid on the future food plant of the caterpillar. In the Silver-washed Fritillary, for example, the female flutters around the base of an oak tree, treading the ground flora, presumably checking for the quantity and condition of violets, and occasionally flying up to the lower branches, probably to check how close she is to the trunk. She then flies to the trunk, often several metres from the ground and proceeds to lay her eggs singly in crevices in the bark. The following spring the eggs hatch and the tiny caterpillars have to make their way to the foot of the tree, then go in search of the violets on which they feed.

BUTTERFLY ENEMIES

So far the butterfly life cycle and behaviour have been described as if everything goes smoothly. Of course, it often does not, and we might well give humans pride of place as the species most responsible for that (see below). Butterflies are viewed by a variety of predators as material for a tasty meal. Many birds, lizards and amphibians are potential threats. Birds, especially, capture butterflies, but the butterflies do have effective defences.

The first line of defence is camouflage. Even brightly coloured butterflies often have much more subdued and cryptic colour patterns on the underside than on the upperside. A would-be predator (and human butterfly watcher) will be confused by the sudden 'disappearance' of a butterfly as it settles and closes its wings to match its background. Some species with very bright uppersides can also 'flash' their wings open suddenly when alarmed. A very common alternative is distraction. Many butterflies bear their brightest markings, often in the form of eye-spots, on areas of their wings most distant from their vulnerable body parts. It is quite common to see butterflies with beak-shaped damage to the outer part of the wing, so this strategy presumably works.

There are invertebrate predators, too. Spiders account for many butterfly deaths, as especially newly emerged butterflies get tangled in webs. However, a more sinister threat are the crab-spiders that lie in wait at flower-heads, often perfectly camouflaged and ready to pounce on an unsuspecting flower visitor. Flies, too, take their toll, especially the distinctive robberflies (Asilidae), which spear their helpless victims, inject poison and suck out their body fluids. Larger dragonflies, too, take butterflies as prey.

Perhaps even more threatening to butterfly survival are the numerous parasitic wasps and flies that infect the early stages in the life cycle – including the egg. Many butterfly enthusiasts collect caterpillars and rear adult butterflies from them, and in the past great disappointment has followed from the discovery that the carefully nurtured caterpillars had been parasitized. Often the shrivelled body of a fully grown caterpillar would be seen surrounded by the tiny cocoons of a parasitic wasp, whose larvae had been feeding on its internal organs. However, as interest in ecology and conservation has grown, these 'accidents' have become a source of interest in their own right.

The main parasitoids (the term given to parasitic species that live on the tissues of their hosts and eventually kill them) of butterflies are flies of the family Tachinidae and parasitic wasps belonging to the families Braconidae and Ichneumonidae. In a few cases the adult parasite lays its egg in the egg of a butterfly, and the whole life history of the wasp takes place within the butterfly egg. More commonly eggs are laid in the caterpillar, sometimes just one in each caterpillar, sometimes in large numbers, resulting in larvae

Peacock butterfly with open wings.

Peacock with wings closed.

Glanville Fritillary, victim of a crab spider.

Dead caterpillar with cocoons of an internal parasitoid.

that feed gregariously within the caterpillar. They may leave the caterpillar to pupate, or leave during the pupal phase. There remains a great deal to be learned about parasitism in butterflies, and especially the impact of parasitism on the population dynamics of the host species. Some parasitoids can use a wide range of host species, whereas others are highly specific in their choice of hosts. It may be that the latter sort of host/parasitoid relationship explains the great population fluctuations of the Holly Blue butterfly. An excellent review of current knowledge is Shaw *et al.*, in Settele *et al.*, 2009.

Butterfly Friends

So much for the enemies. Apart from well-meaning conservationists, butterflies do have some friends. Of course, we could include the plants that provide them with food and shelter at all stages in their lives in this category. However, recent research is increasingly revealing the extent of the importance of the associations between butterflies and ants. These associations seem to have evolved particularly in the family Lycaenidae, especially among some coppers and blues. The most thoroughly studied example is that of the Large Blue.

Research efforts by leading lepidopterist Jeremy Thomas, aimed at saving the Large Blue from extinction in Britain, succeeded in unlocking the secret, but sadly just too late to save the last colony of the butterfly. It was already well known that the caterpillars of the Large Blue began feeding on thyme, but later dropped to the ground and were taken into the nests of ants, where they completed their development. What Thomas showed was that several species of red ant take Large Blue caterpillars into their nests, but that they complete their development in large nests of only one species, *Myrmica sabuleti*. In turn, this species of ant does less well in ungrazed grassland with longer sward. The decline of grazing, and especially the loss of rabbits, on the habitats of the Large Blue had been altering the population ratios of the different ant species, reducing the chances of a Large Blue caterpillar being found by the right ant species. Due to this knowledge and appropriate management, the Large Blue has been successfully reintroduced to several of its former sites in England.

It is now known that close associations with ants are very widespread among Lycaenid butterflies. The European relatives of the Large Blue (the Alcon Blue, and probably also Scarce and Dusky Large Blues) also have close ant associations. However, it seems that the connections are much more widespread. The caterpillars of many blue species can emit pheromones and adopt behaviours that attract ants to them, and also provide 'rewards' to ants in the form of sugary solutions secreted by specialized glands. It has even been discovered that

One of the sites where the Large Blue has been successfully introduced into Britain.

both caterpillars and pupae can emit sounds that mimic the sound communication of the ants. Although this seems like a reciprocal relationship, at least in the case of the Large Blue it is highly exploitative, as the diet of the Large Blue caterpillar, once it has been taken into the nest, consists entirely of ant larvae, with as many as 1,200 being consumed by a single caterpillar, according to Thomas.

BUTTERFLY ECOLOGY

Each species has its own place in a complex web of interactions with other animal and plant species (not to mention the micro-organisms), which together are conditioned by, and may to some extent modify, local soils, climates, seasonality, temperatures, rainfall and so on. Since humans began to have major impacts on the animal and plant communities across Europe, the butterflies that continue to survive here do so largely because they have adapted to, or were already 'preadapted' to, habitats modified by human activity, or because their natural habitats have been relatively unaffected by humans. This latter condition applies to much of the High Alpine and Arctic regions that have been relatively intractable to forestry, intensive agriculture and urbanization, though tourist infrastructure presents a threat in some localities, and climate change a more serious and generalized challenge. Where, alternatively, agriculture, forestry and urban development have radically transformed great swathes of the European landscapes over many centuries, the current challenge to the conservation of wildlife, and butterflies in particular, is the impact of new, more intensive management regimes that have rapidly displaced the older-established, more extensive patterns of management that incidentally sustained the habitats of the locally adapted European butterfly fauna.

Arable monoculture, with semi-natural habitat in the background threatened by urban development.

WOODLAND

Much of lowland Europe at moderate latitudes would have once had extensive woodland cover, with many centuries of localized and small-scale modification by human and animal activity. Woodland was used for subsistence activities such as grazing and collection of firewood, with coppice management providing supplies of wood for fences, handles for agricultural implements and many other uses. The result was a dynamic patchwork with different stages of succession after clearance, including open glades, wood pasture, forest rides and edges. This allowed for the formation of an assemblage of butterflies associated with open deciduous woodland. Some, such as the Purple Emperor and Poplar Admiral, are associated with high forest canopy. Others, like Pearl-bordered and Heath Fritillaries,

A ride in mature woodland, habitat for Silver-washed Fritillary and Purple Emperor.

A woodland glade, habitat for the Heath Fritillary.

Shrubby woodland edge, habitat for the Black Hairstreak.

Damp, open woodland, habitat for the Scarce Fritillary.

Woodland-edge habitat for the Woodland Brown and Large Blue, Sweden.

relied on open, early succession habitats in the woodland. These would be temporary, and since the butterflies have relatively poor powers of dispersal, especially across shady woodland, it was essential for them that patches of open woodland were close together, and linked by connecting habitat.

It is also now recognized that even the butterflies of the high canopy actually require diverse woodland structures. Often habitat for courtship and mating is different from that used for feeding on nectar from woodland shrubs, from the location of suitable host plants for the caterpillars and, again, from the requirement on the part of males of some species to imbibe salts from the woodland floor, from dung or carrion.

Traditional woodland management that sustained the varied conditions for what we recognize as woodland butterflies has been displaced over most of Europe by much more intensive forestry management, often involving replacement of deciduous woodland with dense plantations of fast-growing conifers, or by large-scale clear-fell. The resulting woodland structure is uniform over large areas, eliminating the small-scale diversity necessary to meet the habitat requirements of most species. Alternatively, many small woodland patches were abandoned as management became uneconomical, or were converted to game rearing. Again, the reversion to shady, high-canopy woodland has eliminated the necessary mix of open, sunny and sheltered sites within mature woodland. As a result, many formerly quite common woodland species are now considered threatened both in individual countries and across Europe as a whole.

TEMPERATE GRASSLAND

Over much of lowland Europe, a mixture of woodland and traditionally managed grassland was characteristic of the farmed landscape. Grassland included both pasture for grazing animals such as goats, sheep, cattle and pigs, and meadows, which were cut for hay as animal fodder. These broad patterns of land use would also have been interspersed with both arable agriculture and horticulture. With relatively low stocking levels, pastures would have developed an uneven sward structure, with hoofprints and other disturbances making for patches of bare ground and openings for broadleaved plants to grow. Meadows, too, were frequently flower rich, with a diversity of soil types and levels of soil moisture, and would often have been adjacent to other habitats such as orchards, woodland edges and waysides. Again, a distinct assemblage of European butterflies can be associated with these features – several species of copper, the Brown Argus, Common, Adonis and Chalkhill Blues, among many others. Perhaps half of all European butterfly species occupy such grassland habitats.

The spread of mechanized, high-intensity grassland management regimes has eliminated most of the resources required by this range of butterflies. Flower-rich meadows have been ploughed up and reseeded with dense swards of vigorous, high-nutrient grass species, eliminating many of the flowering plants needed as nectar sources, or larval host plants required by grassland butterflies. In other areas suitable for profitable arable agriculture, grassland has been replaced by monocultures of cereals, oil-seed rape and other cash crops. Yet elsewhere, marginal grassland has simply been abandoned with consequent succession to tall, cool swards of vigorous grasses, and then scrub and secondary woodland. In damp grassland that held some of the most habitat-specialist butterflies (such as Scarce and Dusky Large Blues, and the Alcon Blue), drainage has been the major threat, alongside late-summer grazing that eliminated the flower-heads of the greater burnet on which two of these butterflies lay their eggs.

Chalk downland, southern Britain, habitat of Adonis and Chalkhill Blues and Silver-spotted Skipper.

Flower-rich meadow, Sweden, habitat of Northern Chequered Skipper and Geranium Argus.

Upland limestone grassland, northern Britain, habitat of Northern Brown Argus.

Lowland wetland, habitat of Cranberry Fritillary.

Wetland habitat of Scarce and Dusky Large Blues.

WETLANDS

It is conventional to make a general distinction between two sorts of wetland habitat. The first are the low-nutrient bogs of cooler northern or high-altitude climates, characterized by sphagnum moss and plants such as cotton grass, bilberry, cranberry and bog rosemary. As these habitats are widespread in Arctic and subarctic regions, they are mentioned later. The other sort of wetland is typically high in nutrients and dominated by reeds and sedges, along with plants in the dock and parsley families that are important for several scarce wetland species, such as the Large Copper, Swallowtail, Alcon Blue and Violet Copper in the northern parts of their range. These species are threatened by a natural process of drying out and succession to woodland, and by direct drainage or the indirect effect of a lowered water table in surrounding agricultural land.

HEATHS

Heathland is the result of clearance of areas of light woodland on poor soils, as much as 4,000 years ago. The resulting vegetation is dominated by various species of heather, and traditional management included grazing and fuel collection with regular controlled burning. Heaths are especially important for many species of ant, bee and wasp, but they also sustain an assemblage of butterflies, notably the Silver-studded Blue, Grayling and, where there is some development of gorse or broom scrub, the Green Hairstreak. No longer seen as economic or of scenic value, heathland has been widely used for urbanization or has been left unmanaged, with resulting loss of its value for butterflies. The loss of heathland area combined with the fragmentation of remaining patches has resulted in serious declines of the heathland specialist species.

Lowland heath in eastern England, habitat of Grayling and Silver-studded Blue.

THE FARMED COUNTRYSIDE

Despite justified alarm at the loss of habitat resulting from agricultural intensification, there remain within the farmed landscape in most parts of Europe some residual butterfly habitats. These usually persist in marginal land, such as that on roadside verges, alongside railway tracks, on steep hillsides, around farm buildings, around unmanaged copses, on flood defences, around remaining hedgerows and so on.

These habitats are utilized by migratory species such as the Red Admiral, Painted Lady, Large White and Clouded Yellow, but also provide the resources for resident species such as Large, Small and Essex Skippers, and the Green-veined White, Orange-tip, Common Blue, Small Tortoiseshell, Peacock, Comma, Meadow Brown, Gatekeeper and Small Heath. These residual habitat patches may be supplemented by agri-environmental schemes that fund ecological conservation within commercial farming landscapes. There is evidence that they have benefited the butterflies (and birds) of the wider countryside, but conservation organizations have argued that they have the potential to be more effective (Butterfly Conservation 2015).

Hedgerow, habitat for Gatekeeper, White-letter Hairstreak and Purple Hairstreak.

Roadside verge, Scotland, habitat for Chequered Skipper.

Flower-rich former railway cutting, habitat for Small Blue and Duke of Burgundy.

Conservation field-margin, habitat for Large, Small and Essex Skippers.

A *flowery domestic garden, with Holly Blue, Small Tortoiseshell, Small White and others.*

A *'brown-field' site, habitat for Grayling, Common Blue, Brown Argus and Wall Brown.*

Unmanaged marginal land on the Thames Estuary, southern England.

Restored urban green space, former arable land.

URBAN HABITATS

As the wider farmed countryside has become more wildlife hostile, the attention of many conservationists has turned to a fuller recognition of the value of urban habitats. These include a patchwork of urban and suburban gardens, sports facilities, allotments, informal public green spaces, river margins, roadside verges, railway tracks, canal tow-paths, brownfield sites, churchyards, cemeteries, bridleways and footpaths. This medley of habitats has usually arisen contingently as an unplanned outcome of historical patterns of development, public provision, defensive campaigns by local residents, industrial decline and so on. Increasingly, however, conservationists, together with sympathetic local authorities, have begun to recognize the value of these diverse wildlife 'oases', and to include them in more holistic and integrated planning of green spaces, together with links between them. Recording the species present can then lead to appropriate management plans being devised and implemented. Most of the butterflies that survive in the wider farmed countryside can also thrive in urban habitats, but some of the rarer and more threatened 'habitat specialists' can be provided for, too.

DISTRIBUTION

Europe spans a huge climatic range from the Mediterranean zone (35–45º N) through to more than 70º N in the north of Norway, with much of northern Sweden, Norway and Finland lying within the Arctic Circle. The butterflies of Europe can be roughly divided into four groups in terms of their geographical distribution within Europe.

- One group is primarily associated with the hot, dry scrubland and hardwood forests of the Mediterranean. These species are often confined to areas to the south of the Pyrenees, Alps and eastern European mountain ranges, but some are strong migrants, and arrive regularly or only occasionally in northern Europe.
- A large group is widespread throughout southern and central Europe, reaching its northern limit in Belgium, the Netherlands or the Baltic states.
- The third group is widespread throughout Europe, including Scandinavia and/or Ireland and Britain, though these butterflies are usually more localized in the northern parts of their range.
- A fourth group is adapted to long, cold winters and short summers, and is found in northern Scandinavia and/or at moderate to high altitudes in the European Alps.

The first of the above groups is represented here by species such as the Clouded Yellow, Painted Lady, Long-tailed Blue and Bath White. The second group includes the Mallow Skipper, Red Underwing Skipper, Olive Skipper, Large Chequered Skipper, Sooty Copper, Purple-shot Copper, Blue-spot Hairstreak, Weaver's Fritillary and Knapweed Fritillary. Included in this group can be other species whose northern range extends into southern Scandinavia, such as the Purple Emperor, White Admiral and Niobe Fritillary. Some of these species have recently been extending their range northwards, possibly in response to climate change. However, others, especially habitat specialists, have become more scarce and, in some countries, have become extinct, at the northern limit of their range. This is true of Scarce and Dusky Large Blues, for example.

The third group, the near-ubiquitous species, includes the Swallowtail, Wood White, Orange-tip, Green-veined White, Small Copper, Common Blue, Mazarine Blue and Small Tortoiseshell. Although these species can be found at almost all latitudes, they are usually more localized in the northern parts of their range. This is due to a tendency for species to become more habitat specialist close to the limits of their range. The Swallowtail is an example of this. It is often found in dry scrubland and hillsides in southern localities, but in Britain it is a species of East Anglian wetlands, relying on just one larval food plant, milk parsley *Peucedanum palustre*. Interestingly, migrant Swallowtails from the European mainland are increasingly seen in southern England, so the UK may eventually have two distinct subspecies of this butterfly.

The fourth group, comprising the Arctic-Alpine species, is perhaps the most interesting of the species that occur in northern Europe. The butterflies of high altitudes in the Alpine range are relicts of the retreat of the ice at the end of the last ice age, and were presumably once connected to their relatives that are now isolated in the High Arctic. In some cases, as in the Mountain Fritillary, Alpine Blue and Dewy Ringlet, the same species persist in

Arctic tundra, habitat of the Arctic and Polar Fritillaries and the Arctic Grayling.

the Arctic and at higher altitudes in the Alps or other mountain ranges much further south.

Other Arctic species are clearly close relatives of Alpine species, but have become distinct over several thousand years of separation. Woodland and Arctic Woodland Ringlets are clearly very closely related, as are the Arctic Blue *Plebejus aquilo*, the Alpine and Pyrenean Glandon Blue *P. glandon*, and the Spanish Glandon Blue *P. zullichi*, of the southern Spanish mountains. These examples offer interesting insights to the study of evolutionary change. Many of these species, also including Polar and Arctic Fritillaries, and the Arctic Grayling, inhabit the Arctic tundra habitat of the far north of Scandinavia and western Norway.

Where higher ground above the tree line gives way at lower altitudes to low-nutrient bogs with cotton grass, bilberry and other *Vaccinium* species, bog rosemary and cloudberry, there are other true northern species such as Frigga's and Freija's Fritillaries, the Arctic

Low-nutrient bog with Frigga's and Feija's fritillaries and the Northern Grizzled Skipper.

Bog in northern Sweden, habitat of the Baltic Grayling.

Arctic landscape, northern Norway, showing transitional zone, habitat of the Northern Clouded Yellow.

Habitat of the Pale Arctic Clouded Yellow and Dusky-winged Fritillary, northern Sweden.

Wetland habitat of the Large Heath butterfly, northern England.

Habitat of the Mountain Ringlet, English Lake District.

Ringlet and the Northern Grizzled Skipper. In the transitional zone, occupied by dwarf birch *Betula nana*, with Arctic milk vetch *Astragalus alpinus* and moss campion *Silene acaulis*, with great luck it may be possible to find the two Clouded Yellows of northern Scandinavia, *Colias hecla* and *C. tyche*. Britain and Ireland stand apart from this general account of butterfly distribution in Europe. They have far fewer species than the near Continent, as is usually the case with islands, but their butterfly fauna is also an outcome of the historical vicissitudes of partial recolonization following the last glaciation, combined with the timing of the loss of land links to continental Europe (see R. L. H. Dennis 1977).

BUTTERFLY CONSERVATION

Over many decades, advancing agricultural intensification has imposed severe declines in many butterfly species over most of Europe. Along the Mediterranean coast and in some upland localities, tourist infrastructure has added to the losses, as has the spread of urban settlements. It may be that in more remote parts of northern Europe, especially where altitude combines with climate to limit the impact of agriculture, butterflies have been less affected by human impacts on their habitats. However, even here there is evidence that climate change is subtly affecting the populations of some of the cold-adapted species in the southern parts of their range, and at lower altitudes. There is also some evidence that climate change may be negatively affecting the populations of some other species by allowing them to rear a third generation in late summer, too late for the completion of their life-cycle.

However, over large parts of lowland northern Europe public alarm at the decline of familiar butterfly species has led to the formation of organizations dedicated to butterfly conservation, as well as to serious scientific study of butterfly ecology. Many of the most threatened species have acquired official designation, with the requirement to protect their habitats. For the butterflies of the wider countryside, amateur, or 'citizen science', initiatives have produced detailed evidence-based knowledge of the conservation status of many species, and the development of associated proposals for their protection (see the excellent *The State of the UK's Butterflies 2015*, published by Butterfly Conservation, as an example of such efforts). Over time, the emphasis has shifted from protection of specific sites as nature reserves to protect habitat specialist species, and from small-scale agri-environmental measures, to conservation on a landscape scale. The aim is to have within a larger landscape a mosaic of suitably managed and interlinked habitat patches. This allows for gene flow between subpopulations, and for recolonization if a target butterfly becomes extinct in any of its included sites.

Since butterflies are usually quite readily identified in the field, there is a great opportunity for amateur butterfly enthusiasts to link up with the local branch of a conservation organization and take part in monitoring local butterfly populations, or to help with managing a local patch of habitat. In the UK, Butterfly Conservation has local branches that take on such work, and it also hosts a European Interests Group that works with butterfly conservationists from other European countries.

Skippers (Hesperiidae)

Some of the butterflies in this family resemble moths, and they are omitted in butterfly books by some authors. Skippers are small, fast flying, and have relatively wide heads and well-separated antennae. The family includes several genera, the largest of which is *Pyrgus*. These butterflies have dark brown to black uppersides with an array of white spots, and are quite difficult to distinguish from each another. Another distinctive grouping comprises the ginger-orange species that typically rest with their hindwings flat and their forewings slightly raised (genera *Thymelicus*, *Ochlodes* and *Hesperia*). Both groups have average wingspans of 2.5–3cm. Butterflies of the Mallow Skipper group (genus *Carcharodus*) are sometimes a few millimetres larger.

Dingy Skipper ▪ *Erynnis tages*

DESCRIPTION This small, moth-like butterfly is easily overlooked, but close inspection reveals that it does not deserve its dismissive common name. Upperside ground colour is dark brown, but this is varied by lighter mottling that is especially attractive when the butterfly is freshly emerged. Males have scent scales concealed in a narrow pocket formed by up-curled leading edge of each forewing. Otherwise sexes very similar. Underside plain orange-brown, with tiny pale markings around fringe of hindwing. **DISTRIBUTION** Found throughout Europe to as far north as the Baltic states and southern and eastern Sweden, but absent from Finland. Quite widespread in Britain, but local in England and Wales, becoming more scarce northwards into Scotland. Scarce and localized in Ireland. **HABITAT AND HABITS** Typical habitat is patches of sheltered but sunny grassland, with uneven sward height and areas of bare ground, often close to low scrub. These habitats may be in disused quarries, downland slopes and hollows, coastal landslips and wide forest rides, as in the East Anglian Brecks. Colonies are usually quite small and sedentary, and the butterflies spend much of their time basking in sunshine with their wings outspread. Bird's-foot trefoil is a favoured nectar source, and is also the plant on which the female usually lays her eggs. Flight period is May and early June, with a partial second brood in August in some seasons and in the south of its range.

Male

Female

Mallow Skipper ■ *Carcharodus alceae*

DESCRIPTION Upperside has complex pattern of dark grey-brown patches with lighter fawn to orange-brown bands. Small white marks on forewing soon lose their scales. Underside dark brown with sparse paler spots, and rear margin of hindwing irregularly shaped. **DISTRIBUTION** Common and widespread throughout southern and central Europe, reaching as far north as Belgium and northern Poland, but rare to vulnerable in its northern range. Absent from Britain, Ireland and Scandinavia. **HABITAT AND HABITS** Butterfly of hot, dry and flowery habitats, often on marginal land at the edge of cultivation, fallow fields and 'waste' ground. Individuals bask with wings open in the sun, and fly fast from one perch to another. According to latitude and climate there are several broods each year, and the butterfly can be seen as early as March in the south, continuing into September. Caterpillars feed on various species of mallow and related plants.

Male

Tufted Marbled Skipper ■ *Carcharodus flocciferus*

DESCRIPTION Upperside forewing similar to that of the Mallow Skipper (see above), with brown patches in basal area of wing and darker areas towards margin. Paler central area has small white spots that soon lose their scales, leaving translucent patches. Hindwing has a well-marked band of white spots across it, and an irregularly shaped hind margin. Underside grey or grey-brown with white markings, and pale outlining of veins on hindwing. **DISTRIBUTION** Patchy distribution through southern and central Europe, but more widespread in the east, reaching as far north as the Baltic states. Absent from most of north-western Europe, including Britain, Ireland and Scandinavia. **HABITAT AND HABITS** Occurs in warm, dry, flower-rich grassy areas, often neglected cultivation, roadside verges and other marginal land, as well as woodland edges and sheltered hillsides. Butterflies take nectar from clovers and other meadow flowers, and males also perch on flower-heads ready to intercept passing females. In most places there are two generations each year, mainly in May–June and July–August. Caterpillars feed on woundworts (*Stachys* species) and other plants in the deadnettle group, such as black horehound *Ballota nigra*.

Red-underwing Skipper ■ *Spialia sertorius*

DESCRIPTION Upperside dark brown, often with slightly reddish tint. Scattered white markings on both fore- and hindwings, and row of small white spots close to margins.

Underside hindwing distinctive, with red ground colour, large white spots and pale veins. **DISTRIBUTION** Species of south-western and central Europe, becoming rarer towards its northern limit in Luxemburg and Belgium. **HABITAT AND HABITS** Found in dry, flowery, grassy places open to the sun's rays. Often occupies quite small areas of neglected land by the coast, track sides, open scrub or fallow. This small skipper is easy to overlook as it flies fast and low between perching, or settling on stones or bare ground. There are two generations each year, with flight periods in April–June, and again in July–August. The principal food plant of the caterpillars is salad burnet *Sanguisorba minor*.

Male

Mating pair

Large Grizzled Skipper ■ *Pyrgus alveus*

DESCRIPTION Male upperside dark grey with lighter suffusion, and with pattern of white spots on forewing that is characteristic of the *Pyrgus* group of skippers. Usually just two white spots below central (discal) spot (three in *P. andromedae*), and inner margin of central white spot on underside hindwing is straight (projects inwards in *P. andromedae* and *P. centaureae*). Pale markings on upperside hindwings vaguely defined and faint (well defined and bright in *P. malvae*). **DISTRIBUTION** Present on main European mountain ranges, but becomes more localized in northern Europe. Quite widespread in southern Finland, but more localized and probably declining in southern Sweden and Norway. Absent from Ireland, Britain, the Netherlands and Denmark.

Underside

HABITAT AND HABITS Occurs on open ground with flowers, often rocky slopes in mountains and on the coast (in eastern Sweden). Flight period is from middle of June to mid-August, depending on locality, and the eggs are laid on rock-rose (*Helianthemum* species).

Male

Alpine Grizzled Skipper ■ *Pyrgus andromedae*

DESCRIPTION Distinguished from other skippers in the genus *Pyrgus* by two features. Inner margin of underside hindwing has pale spot and dash forming an '!' mark. On

upperside forewing there are three narrow white spots below the central white spot (discal spot). Also, the paler markings on upperside hindwing are not sharply defined. **DISTRIBUTION** Very disjunct distribution, with a southerly range in the Pyrenees, Alps and some of the mountains of eastern Europe, as well as occurring locally in the mountains of central and northern Scandinavia. Absent from the rest of northern Europe, including Britain and Ireland. **HABITAT AND HABITS** In northern Europe inhabits mountain slopes, usually with bare rocky exposures, which are used for basking. Extracts nectar from flowers such as butterwort (*Pinguicula* species) and has a rapid, low flight. Caterpillars are said to feed on mountain avens *Dryas octopetalla*.

Underside

Male

Oberthür's Grizzled Skipper

■ *Pyrgus armoricanus*

DESCRIPTION Upperside black with white spots on forewing, and indistinct whitish areas on hindwing. Usually a covering of paler scales on forewing, especially on inner half of wing. Underside quite distinctive, having a reddish or yellowish-brown ground colour on hindwing, with paler veins. Largest pale spot close to wing base is usually rounded. **DISTRIBUTION** Species of southern and central Europe, with outlying scattered populations further north in southern Sweden, the Netherlands and Belgium. Absent from Britain and Ireland. **HABITAT AND HABITS** Favours dry, flowery meadows, and flies in two generations each year: from late May to middle of June, and again from late July to end of August. Caterpillars feed on meadowsweet *Filipendula vulgaris* and rock-roses (*Helianthemum* species).

Underside

Male

Safflower Skipper ■ *Pyrgus carthami*

DESCRIPTION Upperside grey-black with dense covering of fine grey hair on basal areas of wings. Wavy line of square white spots on forewing, and dumbbell-shaped central white

spot. Hindwings have obscure whitish markings, notably an arc of elongated white markings close to the margin. Underside hindwing grey-brown with white spots, including an anvil-shaped central spot and distinctive continuous white border. **DISTRIBUTION** Widespread in south-western, central and eastern Europe, reaching as far north as northern Poland and Lithuania, but now extinct in Belgium. Absent from Britain, Ireland and Scandinavia. **HABITAT AND HABITS** Inhabits warm, flowery grassland, often close to scrub or woodland edges, in lowlands to moderate altitudes in mountains. Like most *Pyrgus* species a sun-loving butterfly, often seen basking with open wings. There is one prolonged generation each year, with adults on the wing in June–September, depending on locality. Caterpillars usually feed on cinquefoils (*Polentilla* species), but are said to also use mallows.

Underside

Male

Northern Grizzled Skipper ■ *Pyrgus centaureae*

DESCRIPTION Similar to other *Pyrgus* species, with its dark grey to black upperside with white spots. Males especially have more sharply defined white spots on hindwing than other northern species apart from the Grizzled Skipper (see p. 32). However, the Northern Grizzled Skipper is distinguished from that species by its lack of an outer set of white spots on the forewing. Underside hindwing of this species quite distinctive, having the usual pattern of pale spots on a darker background, but with sharply contrasting white wing-veins. **DISTRIBUTION** Widespread and often common in Sweden and Finland, but more localized in Norway. Absent from Britain and Ireland, and from the rest of northern and western Europe. **HABITAT AND HABITS** Found in sphagnum bogs surrounded by dwarf birch. In sunny weather males fly actively along the edges of a bog in search of females, engaging one another in competitive 'spiral dances', and occasionally settling to take nectar from marsh andromeda *Andromeda polifolia* or cloudberry *Rubus chamaemorus*. In the heat of the day they settle with wings closed, but when the sun goes in they bask with open wings. The eggs are laid on cloudberry and the butterflies fly in mid-June–mid-July.

Male

Female

Underside

Grizzled Skipper ■ *Pyrgus malvae*

DESCRIPTION In Britain this is the sole member of the *Pyrgus* group, so here its black-and-white chequered pattern is distinctive. Where its range overlaps with other *Pyrgus* species, it can be distinguished by the fully developed pattern of clear white spots on the upperside hindwing. Distinguished from the Northern Grizzled Skipper (see p. 31) by extra row of white spots close to border of forewing. **DISTRIBUTION** Widespread in most of Europe to as far north as southern Scandinavia. Widespread in Britain, but local in southern and central England and parts of Wales, and reduced to a scattering of isolated populations

further north. **HABITAT AND HABITS** Inhabits sheltered but sunny patches of rough grassland with bare ground, and both short and longer swards, often in long-disused ex-industrial sites, uncultivated track margins and woodland edges. Although their larval food plants are different, the habitats of this species and the Dingy Skipper (see p. 24) overlap and the two often fly together. Males spend much of their time basking, occasionally flying up to intercept a passing female, or to challenge another male. They fly from as early as mid-April through May and into June. According to place and season there may be a second appearance in August. The most favoured food plant of the caterpillars is wild strawberry *Fragaria vesca*, but several other plants in the rose family, including cinquefoils and bramble, may be used.

Male (below), female (above)

Undersides

Olive Skipper ■ *Pyrgus serratulae*

DESCRIPTION Upperside shares characteristic *Pyrgus* pattern of dark brown with scattered white spots. In this species white spots are small, and hindwing upperside has only a very faint paler area, or may be plain dark brown. Underside hindwing olive-brown or yellowish-brown, rather plain, with square central white spot, and oval white spot near base of wing. **DISTRIBUTION** Quite widespread in central and eastern Europe, but absent from large parts of the Iberian peninsula, Italy and the Mediterranean islands. Northwards it reaches Belgium, where it is considered endangered and has outposts in the Baltic states. Absent from Scandinavia, Britain and Ireland. **HABITAT AND HABITS** Habitat is open, dry, flowery and grassy places, often with patchy scrub. There is one generation each year, in May–July, later at high altitudes and in the north. Caterpillars feed on tormentil and various cinquefoils (*Potentilla* species).

Underside

Female

33

Chequered Skipper ■ *Carterocephalus palaemon*

DESCRIPTION Upperside warm brown with pattern of yellow-orange spots. These are angular in forewing, with a row of smaller spots along wing margin. Hindwing also has a row of small yellow spots on margin, with a group of larger rounded spots occupying rest of wing surface. Underside pattern on hindwing repeats that on upperside but in much more subdued hues. **DISTRIBUTION** Very wide European distribution, including most of Norway, Sweden, Finland and the Baltic states, but absent from Denmark and Ireland. It occurred, sometimes commonly, in wooded areas in the English midlands until the 1960s,

finally becoming extinct by the mid-1970s. In 1939 it was discovered near Fort William in Scotland, and it is now known to be well established in south-western Scotland. **HABITAT AND HABITS** The former English habitats were woodland rides and glades, but in Scotland it seems to be more general, inhabiting roadside verges, and open, scrubby habitats by streams and lochs, as well as broadleaved woodland. Males establish prominent perches, from which they make quick sorties to see off passing males or intercept females. Both sexes take nectar from plants such as bluebell, bugle and speedwells, but they seem especially associated with bluebells. The eggs are laid on various grasses, depending on locality. In Scotland purple moor grass is used. Flight period is from late May to mid-June.

Underside

Male

Northern Chequered Skipper ■ *Carterocephalus silvicolus*

DESCRIPTION Male upperside very striking, with shining golden ground colour on forewings, and darker markings. Hindwings have large, gold-yellow markings. Female upperside very different, resembling that of Chequered Skipper (see p. 34), except that yellow markings on forewing run into one another. Underside pattern a subdued version of upperside.

DISTRIBUTION Occurs in northern Germany, Poland, the Baltic states, southern Finland, southern and western Sweden, and a few localities in Norway. Absent from Britain and Ireland.

HABITAT AND HABITS Favours lowland flowery meadows, roadside verges, and woodland glades and fringes. Males bask with wings open, and take nectar from cranesbill and other meadow flowers. Flight period is from late May to June, and the eggs are laid on various grasses.

Male

Female

Underside

Large Chequered Skipper ■ *Heteropterus morpheus*

DESCRIPTION Very distinctive butterfly with large, black-ringed whitish spots, set in a yellow background, on underside hindwing. Upperside warm brown, with a few

yellowish spots towards apex of forewing.
DISTRIBUTION Rather a patchy distribution in western, eastern and parts of northern Europe. Occurs in the Baltic states, southern tip of Sweden, Belgium, northern Germany and north-eastern Poland. Absent from Britain and Ireland, as well as most of Scandinavia.
HABITAT AND HABITS Habitat is tall, grassy, open spaces and ride margins in woods, or edge of woods, as well as damp meadows. Butterflies are very easy to spot, as they fly with a unique 'bobbing' action over vegetation, but are much harder to approach for photography. Flight period is from late June to early August, and females lay their eggs on various coarse grasses.

Underside

Male

Lulworth Skipper ■ *Thymelicus acteon*

DESCRIPTION Adults resemble Small and Essex Skippers (see pp. 39 and 38), but the upperside colouration is darker. Male upperside yellow-orange variably suffused with grey-brown, and with fine black band of scent scales on forewing. Females have an arc of orange spots across outer part of each forewing. **DISTRIBUTION** Common and widespread in southern Europe, but more local further north to the Netherlands and northern Germany. Absent from Scandinavia, the Baltic states, Denmark and Ireland. In England confined to

coastal habitats close to south coast, mostly in the county of Dorset. **HABITAT AND HABITS** In England inhabits south-facing, uncultivated chalk downland, coastal cliffs and landslip. Here, at its northern climatic limit, it is active only in sunshine. Butterflies are on the wing from the end of June through to late august, and take nectar from a range of downland flowers, including self-heal, cranesbills, clovers and marjoram. The eggs are laid midway up the stems of tor-grass, growing in tall tussocks, so the butterfly is very sensitive to grazing intensity.

Underside

Male

Female

Essex Skipper ■ *Thymelicus lineola*

DESCRIPTION Very similar in appearance to the Small Skipper (see p. 39), and not recognized as a distinct species in England until the late 19th century. Close observation is needed to separate the species: underside of tip of antennae of the Essex Skipper is black, as if dipped in ink, while in the Small Skipper it is usually yellow-orange. Care is needed in identification, as upperside of antennae of the Small Skipper is often darkened. Black line of scent scales in male is much finer in the Essex Skipper, and runs parallel to edge of forewing (oblique in the Small Skipper). **DISTRIBUTION** European distribution extends

into southern Sweden, Finland and Norway. In Britain it has been seen as predominantly a species of south-eastern England, though it is spreading both north and west, and has recently been identified in the south-east of Ireland. **HABITAT AND HABITS** Flies in a wide variety of grasslands, such as sea walls and coastal grazing marshes, as well as roadside verges and embankments. Tends to favour dryer habitats than the Small Skipper, but the two species often fly together. This species adopts a similar pose to that of the Small Skipper when at rest and, like that species, tends to make only short, darting flights. Flight period is also late June–August, peaking slightly later than that of the Small Skipper. The eggs are laid on several species of wild grass.

Male

Head and antennae

Underside

Small Skipper ■ *Thymelicus sylvestris*

DESCRIPTION Upperside bright orange with narrow black border and, in males, oblique black line of scent scales across forewing. Underside colouration more subdued, with grey-green suffusion predominating over pale orange on hindwing. **DISTRIBUTION** Very widespread and usually common throughout Europe except Norway, Sweden and Finland. Absent from Ireland. In Britain it is close to its northern limit, but has been extending its range in recent decades. It now inhabits the whole of England and Wales, and reaches over the Scottish border. **HABITAT AND HABITS** Inhabits a wide variety of grasslands, including coastal flood defences, downland, roadside verges, open woodland rides and edges, and even parks and gardens where wild grasses are allowed to grow tall. Butterflies take nectar from a wide variety of flowers, including bramble, clovers, tufted vetch, bird's foot trefoil and thistles. Much of their time is spent perched on grasses, with forewings part raised and hindwings spread flat. Females lay their eggs in the sheaths of grass stems, usually Yorkshire fog *Holcus lanatus*, and the flight period is mid-June–August.

Head and antennae

Male

Silver-spotted Skipper ■ *Hesperia comma*

DESCRIPTION Upperside in both sexes orange with dark grey-brown borders on forewings, and more extensive darkening on hindwings. In male forewing there is a strongly marked oblique band of scent scales, and the dark border has a cluster of small yellow patches towards the apex. Female forewing has more extensive darkening, and extra yellow patches. Underside is strikingly beautiful, the olive-green of the hindwing being studded with bright white spots. **DISTRIBUTION** Widespread in Europe, becoming more

localized to the north. Occurs widely in southern Scandinavia and then, as a distinct subspecies, in the far north of Norway. In Britain confined to chalk and limestone hills in southern and south-eastern England, and absent from Ireland. **HABITAT AND HABITS** Butterfly of south-facing chalk or limestone downland, usually quite sparsely vegetated and with bare, stony ground that the butterflies use for basking. When at rest they adopt the common pose of this group of skippers, with forewings held obliquely above the wide-spread hindwings. Both sexes take nectar from a wide range of flowers, including scabious, knapweed, marjoram and stemless thistle. Butterfly is on the wing during August, and the eggs are laid on sheep's fescue grass.

Underside

Male

Female

Large Skipper ■ *Ochlodes sylvanus*

DESCRIPTION Upperside very similar to that of the Silver-spotted Skipper (see p. 40), except that dusting of dark scales on wing borders is lighter, so that overall impression is less contrasting. Underside pattern, too, is quite similar, but white markings on hindwing are yellow, not brilliant white. **DISTRIBUTION** Widespread and often common in most of Europe, including the Baltic states, Belgium, the Netherlands, Denmark, southern Norway and Finland, and southern and eastern Sweden. In Britain widespread in England and Wales, reaching as far north as southern Scotland. **HABITAT AND HABITS** Butterfly of rough grassland, often occupying quite small patches of uncultivated field corners, roadside verges, hedge banks and track sides. Males alternate between patrolling and perching in wait for a passing female, and behave territorially with other males. They take nectar from bird's-foot trefoil, meadow vetchling, bramble, creeping thistle and other wild flowers. Flight period is prolonged, lasting from mid-May to late August or even September. Depending on locality, the eggs are laid on cock's-foot grass or purple moor grass.

Male

Female

Underside

SWALLOWWTAILS, FESTOONS & APOLLOS (PAPILIONIDAE)
This is a family of large, often quite spectacular butterflies, some of which have been greatly prized by collectors in the past. Swallowtails have, as their English name implies, tail-like extensions from the rear of their hindwings. Festoons (none of which occur in northern Europe) are quite similar but lack the tails, while the apollos have large, papery white wings, in some species with variable red or orange spots. The hindwings of all species have concave inner margins. The caterpillars often have warning colouration, at least in the later stages of development, and can secrete an offensive odour. Swallowtails and apollos have average wingspans of 6–7cm, with the Clouded Apollo's being rather smaller, at 5.5–6cm.

Scarce Swallowtail ■ *Iphiclides podalirius*

DESCRIPTION Unmistakable large, handsome butterfly, with wide wings tapering to long, trailing tails from hindwings. Forewings pale yellow-white with black streaks, one of which continues onto each hindwing. Outer edge of each hindwing has a series of blue lunules, and one on the inner edge, capped by an orange arc. Underside echoes upperside pattern.
DISTRIBUTION Widespread and often common in southern and central Europe, becoming

more scarce northwards to its current northern limit in the Netherlands. Absent from Britain, Ireland and, except as a rare vagrant, from Scandinavia.
HABITAT AND HABITS Favours areas of human habitation, including urban parks and gardens, orchards, field margins, scrub and light woodland. The butterfly is very graceful on the wing as it glides from flower to flower, and is easily approached as it feeds. There are two generations each year, the first in May–June, the second in August–September. The eggs are laid on shrubs such as hawthorn and blackthorn, as well as various fruit trees (*Prunus* species).

Swallowtail ■ *Papilio machaon*

DESCRIPTION This iconic species is distinguished by its large size, the 'tails' that protrude from its hindwings, and the lovely upperside pattern of yellow patches with darkened wing-veins and dark, wide borders that are studded with brilliant blue patches and orange spots on the hindwing. Underside has similar pattern to upperside, but colours are of more delicate shades. **DISTRIBUTION** Widespread and relatively common species throughout Europe, becoming more thinly distributed in northern Scandinavia. Formerly widespread in wetlands of eastern England, but currently confined to the Norfolk Broads. Absent from Ireland. **HABITAT AND HABITS** British populations belong to a distinct subspecies that inhabits reedy fens and marshes, and lays its eggs on milk-parsley. They are on the wing through June, with a partial second generation in August. Elsewhere in Europe the Swallowtail is found in a wide range of habitats, and can even be seen on mountain tops. The eggs are laid on fennel and other plants in the carrot family. Immigrant butterflies from mainland Europe reach southern England frequently, and are known to breed in some southern counties.

Underside

Apollo ■ *Parnassius apollo*

DESCRIPTION Upperside white with grey suffusion near wing bases and faint grey lines towards outer edges. Several bold black spots on forewings and, on each hindwing, two large, black-ringed, red eye-spots. Scales thinly spread on wings, and outer edges quickly become semi-transparent. Underside similar, but with several small additional red spots close to base of hindwing. **DISTRIBUTION** This iconic butterfly has a discontinuous distribution in Europe, occurring in mountains of south-western and central Europe,

and the Balkan peninsula. Further north, in southern Sweden, Norway and Finland, it is found at lower altitudes. Absent from Ireland, Britain and the rest of northern Europe. **HABITAT AND HABITS** Occupies open, well-drained areas with bare rock ('limestone pavement'), with scattered small trees and shrubs. Males fly patrol routes, contouring trees and shrubs and soaring in the breeze. They mate on the ground, and females lay their eggs on various plants, including juniper. However, the caterpillars (which over-winter in the egg shell) feed on stonecrops (*Sedum* species) and related plants of dry habitats. Butterflies are on the wing in late June–August.

Male

Clouded Apollo ■ *Parnassius mnemosyne*

DESCRIPTION Both sexes are white, with wing veins outlined by black scales, two black spots towards leading edge of forewing, and sometimes one spot in centre of

hindwing. Wide outer borders of forewings very thinly scaled, and appear translucent. Rather stocky body is covered in pale hairs. **DISTRIBUTION** Mainly a butterfly of southern and eastern Europe, with scattered populations further north in the Baltic states, and the south of Sweden, Norway and Finland. **HABITAT AND HABITS** Occurs in damp, grassy areas in forests and on mountain slopes. A distinctive feature of the apollo group of butterflies is that males attach a sheath to the female's abdomen after mating to prevent her from mating with other males. Butterflies are on the wing in May–July, depending on locality, and the eggs are laid on several species of *Corydalis*.

Female

WHITE & YELLOWS (PIERIDAE)

This family includes some of the most familiar butterflies, and they are probably undervalued because of that. Two distinctive groups within the family are, first, the white butterflies of the genus *Pieris* – Large, Small and Green-veined Whites – which are among the most successful species in coping with human-wrought changes. They, like many other butterflies in this family, are closely associated with plants in the cress, or cabbage, family (Cruciferae), including in some cases cultivated species. The other distinctive group are the clouded yellows, all characterized by some combination of yellow with black markings, and favouring plants of the vetch family (Fabaceae) as food plants for their caterpillars. The pattern of white or yellow sometimes with an orange patch on the forewing applies to another widespread European grouping (genus *Gonepteryx*), but there is only one such species in northern Europe (the Brimstone). Most of the species in this family are medium in size, the whites usually averaging 4–4.5cm wingspans, the clouded yellows usually a few millimetres more. The Large White, Black-veined White and Brimstone are somewhat larger, with wingspans averaging 5–6cm. The Wood White is among the smallest in the family, with a wingspan of usually up to 4cm.

Wood White/Cryptic Wood White ▪ *Leptidea sinapis/L. juvernica*

DESCRIPTION Small, delicate butterfly with rounded wings. Upperside white with black or grey marks at tips of forewings. Underside hindwing has pale grey-green, fan-shaped markings. **DISTRIBUTION** The species previously known as *Leptidea sinapis*, the Wood White, has since the late 1980s been recognized to comprise two or three 'sibling' species. There are no reliable features that distinguish them without dissection, so their respective distributions are still not fully established. In most parts of northern Europe the ranges of the Wood White and the Cryptic Wood White overlap, and the two species often fly at the same localities. It seems that British Wood Whites are all *L. sinapis*, while in Ireland most are *L. juvernica*, except in a small area on the west coast, where only *L. sinapis* flies. In Britain the Wood White has a very scattered distribution in southern and central England and Wales. **HABITAT AND HABITS** The Wood White inhabits open, sunny rides in woodland, but also occurs on verges and cuttings on disused railway lines. The males fly with slow wingbeats along rides, in search of females. When the two sexes meet they engage in a distinctive head-to-head courtship. Both sexes take nectar from flowers such as bugle, bird's-foot trefoil and herb robert, or knapweed later in the year. The Cryptic Wood White is more likely to be found in open grassland, but both species lay their eggs on a similar range of plants, most especially meadow vetchling, bird's-foot trefoil and related species in the pea family. Flight period is from late April to middle of June, with a partial second brood from mid-July to late August.

L. sinapsis *feeding*

Orange Tip ■ *Anthocharis cardamines*

DESCRIPTION Male uppperside white, with a black spot in centre of forewing, a wide orange patch covering much of outer portion and a grey-black tip. Underside hindwing mottled olive-green. Female very similar to male, but lacks orange on forewing. **DISTRIBUTION** Widespread and usually common throughout Europe, becoming more localized in northern Scandinavia. Common throughout Britain and expanding range in Scotland. **HABITAT AND HABITS** Habitat is damp, unimproved meadows, but as these have been lost it is more often seen along hedgerows, woodland edges and verges, or in damp hollows or river banks where the food plants of the caterpillars grow. Males are a

delightful sight in spring, as they fly tirelessly along a hedgerow in search of a mate, occasionally taking nectar from cuckoo flower or herb robert. Females are less obvious, and very well camouflaged by their mottled undersides. They lay their eggs on flower-heads of cuckoo flower, garlic mustard and other plants in the cress family (Cruciferae). Usually only one egg is laid on each flower-head, as the (cannibalistic) caterpillars feed on the seed pods, and there is often barely enough to nourish one. Butterflies are on the wing in April–June, with an occasional second brood later in the year.

Male

Female

Male showing underside

Black-veined White ■ *Aporia crataegi*

DESCRIPTION As its name implies, this butterfly is white with wing veins marked in black. There are patches of grey scales towards the edges and tips of the forewings, but the butterfly soon becomes worn, and the semi-transparent wing membrane is exposed.

DISTRIBUTION Widespread across mainland Europe, including southern Sweden and Finland, scarce in southern Norway and extremely localized further north in Scandinavia.

Formerly established in England, but with wide fluctuations in population. In Kent it was even treated as a pest of fruit orchards in years of abundance, but it became extinct at its last known localities in the early 1920s. **HABITAT AND HABITS** Sometimes found in large aggregations and frequently roosts communally, draped over a plant stem. Takes nectar from flowers such as geraniums, cow wheat and thistles, and is on the wing from late May to mid-July. The eggs are laid on a wide range of shrubs and trees, including hawthorn, mountain ash and blackthorn, as well as cultivated *Prunus* and other fruit trees.

Underside

Large White ■ *Pieris brassicae*

DESCRIPTION Sometimes called the Cabbage White, and resented by gardeners as a pest species, the Large White is a very striking insect. Male upperside pure white with grey-black apex to each forewing, and small black mark on leading edge of each hindwing. Female similar, but with two black spots and black dash on each forewing. Underside forewing white with yellowish tip and black central spots, and underside delicate yellow, sometimes suffused with darker scales. Can be recognized on the wing by large size and

more powerful flight than that of the other white butterflies. **DISTRIBUTION** Common and widespread throughout Europe except northern and central Scandinavia. Resident throughout Britain and Ireland, but numbers are augmented each year by migrants. **HABITAT AND HABITS** Due to its strongly migratory habit, the butterfly can be seen in almost any habitat, urban or rural. Females lay their eggs on a wide range of plants in the cress family (Cruciferae), but they favour the larger cultivated brassicas. A female may lay 40–100 eggs in a cluster on a cabbage plant, with the resulting caterpillars almost completely defoliating it. Treated as a pest by gardeners, and suffering both parasitism from a tiny wasp and the depredations of a viral disease, it is surprising that the species remains so common. Butterflies appear in May and continue in two or more broods throughout the summer into the autumn.

Underside

Male

Female

Green-veined White ■ *Pieris napi*

DESCRIPTION This small and delicate species can easily be overlooked, but its discreet beauty rewards closer inspection. Male upperside white with faint grey veins, dark tips to forewings and central spot. Markings are usually more prominent in later broods. Females have more strongly outlined veins, two black spots and a dash on each forewing, and more extensive grey dusting around wing-tips. Underside is particularly beautiful, the hindwing being lemon yellow with green outlining to the veins. In northern Scandinavia there is a subspecies in which the females have much more extensive dark suffusion over the upperside. **DISTRIBUTION** Occurs in numerous locally adapted forms throughout Europe, including northern Scandinavia. Common and widespread throughout Britain and Ireland. Much less of a wanderer than Large and Small Whites (see pp. 48 and 50), and much more discerning in its habitat preferences. Can be found in humid, sheltered grassland, often close to streams or rivers, and sheltered by scrub or woodland edge. Damp hedgerows and verges are also often used. **HABITAT AND HABITS** Butterflies emerge in April and can be seen until late June, taking nectar from buttercup, cuckoo flower, bluebell and other spring flowers. The eggs are laid on various wild plants in the Cruciferae family, such as cuckoo flower, hedge mustard and garlic mustard. These result in a more numerous summer brood, at its peak in July.

Female, Scandinavia

Underside

Male

Female

Small White ■ *Pieris rapae*

DESCRIPTION Smaller than the Large White with ground colour slightly off-white to creamy. Male upperside has greyish tip to forewing and central spot, while hindwing has small dark spot on leading edge. Female has additional black spot and faint dash

Female

Male

on each forewing. Underside pattern similar to that of the Large White (see p. 48). Summer broods tend to be more strongly marked, resembling the Large White more closely. **DISTRIBUTION** Common throughout Europe, including the Baltic states and southern Scandinavia. Further north in Sweden, Norway and Finland it is very scarce or absent. Widespread throughout Britain and Ireland. **HABITAT AND HABITS** Ranges over both urban and rural habitats, being found in parks and gardens as well as along hedgerows, verges, woodland edges and uncultivated grassland by the coast. According to latitude and climate may be on the wing as early as March, but population peaks in May. There is then a second emergence in late June or July, and sometimes a later one in the autumn. Numbers are usually augmented in summer by migrants. The butterflies do not form local colonies, but range across the country, males searching for potential mates, while females seek suitable plants on which to lay their eggs. Like the Large White they use cultivated plants in the cabbage family, but wild plants in the same family are also used, especially garlic mustard and charlock.

Bath White/Eastern Bath White ■ *Pontia daplidice/P. edusa*

DESCRIPTION These butterflies are treated as distinct species on scientific grounds, but are not distinguishable in the field. Upperside white with bold black markings towards apex of forewing, and central black spot bisected by white vein. Males have reduced black markings, especially on hindwings. Underside hindwing has regular mottled pattern of olive-green and white, with fine yellow veins noticeable across white areas.
DISTRIBUTION Both species are common residents in the south, but reach as far north as southern Britain, Ireland and Scandinavia as migrants, or descendants of migrants in variable numbers from year to year. As their common names imply, they occupy western and eastern zones of Europe, respectively. **HABITAT AND HABITS** Can be seen almost anywhere as they are strongly migratory, but they breed in places where there is thin, poor soil. This includes abandoned cultivation, worked-out quarries, embankments, fallow land and brownfield sites where the caterpillar's food plants thrive. There are 2–4 generations each year, so the butterflies may be on the wing from March to September or October. Migrants are more often seen in mid- to late summer, but very few reach the shores of southern Britain. Eggs are laid on weld and mignonette (*Reseda* species) as well as on several wild plants in the cress family (Cruciferae) such as hedge mustard and rocket.

Underside

Male

Female

Clouded Yellow ■ *Colias crocea*

DESCRIPTION Upperside in both sexes mainly rich orange-yellow, with black borders to wings. Borders broken by row of yellow spots in female. Dusting of darker scales towards bases of hindwings, especially in females. Underside yellow with greenish tint on hindwing and forewing margins. Central black spot in underside forewing, and row of dark markings towards margin of both fore- and hindwings. There is a fairly frequent form of the female

Female, f. helice

(f. *helice*) that is greenish-white on the upperside forewings and stongly suffused with dark scales on the hindwings. **DISTRIBUTION** Resident species in southern Europe, but strongly migratory, and reaches much of northern Europe in most years, occasionally in very large numbers. **HABITAT AND HABITS** Migrant males are seen flying fast over open ground, perching occasionally to take nectar from a wide range of flowers, while females are more often seen searching out suitable plants for egg-laying. These include bird's-foot trefoil, lucerne, vetches and other plants in the pea family. Butterfly breeds in continuous broods through spring and summer, but cannot survive the winter in northern Europe.

Northern Clouded Yellow ■ *Colias hecla*

DESCRIPTION Upperside ground colour bright orange in both males and females. Both sexes have black spot in middle of each forewing, and males have broad black border to both fore- and hindwings. In females black border is broken by row of yellow patches. Underside hindwing grey-green, with faint yellowish areas towards wing border, and central pale spot with red border. Forewing yellow with central black spot, and dark upperside borders show through with back-lighting from the sun. **DISTRIBUTION** In Europe this Arctic butterfly occurs only in the far north of Norway, Sweden and Finland, within the Arctic Circle. **HABITAT AND HABITS**

Characteristic habitat is mountain slopes, just above the tree-line, where males begin their fast, tireless flights in search of females. Early in the morning they can be seen taking nectar from patches of moss campion. Females lay their eggs on low-growing vegetation in the transitional zone between birch woodland and open heath, where the main food plant of the caterpillar, alpine milk-vetch *Astragalus alpinus*, grows among the dwarf birches. Between episodes of egg-laying the female settles and becomes torpid. Butterflies are on the wing from mid-June to late July.

Female

Male

Pale Clouded Yellow/Berger's Clouded Yellow

■ *Colias hyale/C. alfacariensis*

DESCRIPTION These species are very difficult to distinguish, and it is likely that many reports have confused them. Both have pale yellow uppersides with dark wing borders. However, they always settle with wings closed. Underside in both species pale yellow with red-ringed central white spot in hindwing, and central black spot in forewing. Row of

Male, C. hyale

small, dark markings around margin of hindwing. The Pale Clouded Yellow is said to have a more curved leading edge to the forewing and a more pointed apex, but the differences are slight and variable. **DISTRIBUTION** The Pale Clouded Yellow is an uncommon but regular immigrant in southern Scandinavia and Britain, with breeding occasionally recorded, while Berger's Clouded Yellow has a more southerly distribution and is only rarely recorded as a migrant in northern Europe. **HABITAT AND HABITS** The Pale Clouded Yellow lays its eggs on clovers, lucerne and other plants in the pea family, while Berger's Clouded Yellow seems more closely associated with horseshoe vetch. Migrant individuals arrive during the spring, with their descendants being seen in August and September.

C. alfacariensis, *pair*

Danube Clouded Yellow ■ *Colias myrmidone*

DESCRIPTION Upperside of male deep yellow-orange with a black spot in middle of forewing, and broad black borders. Female upperside paler and black borders enclosing lemon-yellow spots. Male underside hindwing yellow, slightly suffused grey and with orange-ringed central white spot. Orange ground colour of upperside shows through with back-lighting. Female underside paler, with upperside markings faintly visible. **DISTRIBUTION** Its European distribution is closely related to the Danube river system, becoming more widespread in the east into western Asia. In Europe a rare and rapidly declining species, with (former?) outposts in the southern Baltics. **HABITAT AND HABITS** Inhabits lowland grassland with scattered shrubs and light woodland. Adults take nectar from available wild flowers, and males drink from damp ground, always with wings closed. The caterpillars are said to feed from various species of broom, and there are two flight periods each year, May to June, and July to September.

Male underside

Moorland Clouded Yellow ■ *Colias palaeno*

DESCRIPTION Upperside very pale yellow in males, greenish-white in females, with clearly defined black borders (sometimes with small whitish markings within borders in females). Underside yellow-green with dusting of darker scales and white spot in centre of hindwing, and black-bordered white spot at centre of forewing. **DISTRIBUTION** Familiar species in the Alps, becoming more localized further north in Europe, but widespread and fairly common in lowland southern Sweden, Norway and Finland, and extending to the far north. Absent from Britain and Ireland. **HABITAT AND HABITS** In southern Sweden inhabits bogs and marshes where the food plant of the caterpillar, bog whortleberry *Vaccinium uliginosum*, grows. Males fly fast and low over bogs, settling only in dull weather. They are on the wing from mid-June to mid-August. In Arctic Scandinavia they fly in mountainous areas, and are on the wing a little later in the year.

Pale Arctic Clouded Yellow ■ *Colias tyche (nastes, werdandi)*

DESCRIPTION Always settles with wings closed, but can be readily identified by subtle grey-green hindwing underside of female, faintly yellow-tinted in males. Forewings have grey borders, but these are much less densely marked than in other clouded yellow species. **DISTRIBUTION** Restricted in Europe to the mountains of northern Sweden, Norway and Finland. **HABITAT AND HABITS** Within its range can be found in a variety of habitats, including heaths, bogs and mountain slopes up to 1,000m. Males 'patrol' flowery slopes, darting between patches of moss campion *en route*. Caterpillars feed on milk-vetches (*Astragalus alpinus* and *A. frigidus*).

Female

Brimstone ■ *Gonepteryx rhamni*

DESCRIPTION Male upperside bright lemon-yellow, with small red central spot in each wing. Female similar but pale greenish-white. Undersides pale yellow or greenish-white, according to gender, with prominent white veins, especially on hindwings. Wing shape is perhaps the most distinctive feature: hooked tip to forewing, and small 'tail' on hindwing. **DISTRIBUTION** Occurs throughout Europe, but with only very scattered reports from northern and western Scandinavia. Widespread throughout Ireland and Britain except Scotland. **HABITAT AND HABITS** Although widespread, the Brimstone is absent or scarce within some parts of its range, as it is dependent on the availability of the food plants of its caterpillars, buckthorn and alder buckthorn. Butterflies emerge in early July and take nectar from flowers such as thistles, knapweeds, scabious and buddleia, often in gardens. They soon find a sheltered spot (often among ivy leaves), where they spend the winter. In spring they emerge from hibernation, when the males fly considerable distances in search of females, feeding from bluebells and other spring flowers. After courtship and mating the females seek out buckthorn shrubs in sunny, sheltered spots and lay their eggs. A few worn individuals may still be flying when their offspring emerge.

Female

Male

> **Metalmarks (Riodinidae)**
> There is only one European species in this large, mainly tropical family. The wingspan
> of the Duke of Burgundy is approximately 2.5–2.7cm.

Duke of Burgundy ■ *Hemearis lucina*

DESCRIPTION Orange-and-black upperside pattern reminiscent of some fritillary
butterflies, though it is much smaller than any northern European fritillary. Underside
paler orange with scattered darker markings, and two rows of large, oval to oblong white
markings across hindwing. **DISTRIBUTION** Mainly a species of central Europe, absent
from the far south, and increasingly localized northwards to the Baltics and southern
Sweden. In Britain it survives in parts of south-central England, Kent, Cumbria and North
Yorkshire, but has declined greatly in recent decades. **HABITAT AND HABITS** There are
two distinct habitats for this butterfly. One is scrubby downland, usually north-facing, with
cowslips; the other is open woodland with primroses. The butterflies emerge from late
April, with numbers reaching a peak in the latter part of May. Males are very territorial,
occupying a prominent perch from which they fly up to challenge passing insects, or to
waylay potential mates. The eggs are laid on the undersides of the leaves of *Primula* species,
cowslip or primrose, depending on the habitat. In woodland the butterflies thrive following
coppicing, or clear-fell in the case of some conifer woods, but need to find other suitable
habitat as the canopy closes over. Cessation of coppicing over most of Britain has led to
many local extinctions, and the future of the butterfly here seems to depend on active
conservation measures.

Underside

Female

HAIRSTREAKS, COPPERS & BLUE (LYCAENIDAE)

The very large and diverse family Lycaenidae includes three distinct groups, all well represented in northern Europe.

The hairstreaks are so-called because of the fine line (or series of dots) running across each underside hindwing. They are mostly species of woodland, being strongly associated with trees or shrubs. Most species settle with their wings closed.

The coppers include some of the most beautiful of all butterflies, with the males usually the more spectacular sex. The caterpillars of most species feed on sorrels, bistorts and other plants in the dock family (Polygonaceae).

The third and largest group, the blues, is itself very diverse. Some species are not blue at all, while in others the males are blue and females mainly brown, or partly blue. The caterpillars of most species feed on vetches, trefoils and other plants in the pea family (Fabaceae), and some have close associations with ants. Most species are small, with average wingspans of 2.5 to 3cm. The Small Blue is smaller, with a wingspan of only 2.2cm, while the large Copper, Purple-edged Copper and the Large Blues are a little larger at 3–3.5cm.

Green Hairstreak ■ *Callophrys rubi*

DESCRIPTION Upperside in both sexes plain dark brown (becoming rusty-brown with age), and with small pale patches of scent scales on male forewing. Underside green with row of linear white marks close to margin of hindwing. This is sometimes reduced to just one or two spots, or it may be complete and continued onto forewing. Wing margin irregularly shaped at tail end. **DISTRIBUTION** Widespread throughout Europe, reaching northern Scandinavia. Widespread in Britain and Ireland, but often quite localized, especially in eastern England. **HABITAT AND HABITS** Inhabits wide range of scrubby habitats, including lowland heaths and moors, sheltered downland hollows, wide woodland rides, and sometimes garden hedges and uncultivated scrubland in urban areas. Males are particularly evident, as they perch on shrubs ready to fly up and challenge passing males, or to court a female. The eggs are laid on a wide range of plants including bilberry, gorse, broom, bird's-foot trefoil, bramble, buckthorn and dogwood, according to habitat. The chrysalis is formed on the ground and is attended by ants. Butterfly emerges from mid- to late April, reaching a peak in late May.

Male, well-marked form

Brown Hairstreak ■ *Thecla betulae*

DESCRIPTION Male upperside dark brown with small orange patch next to dark central spot in forewing, and small orange flashes close to 'tail' on hindwing. Female upperside has much more extensive orange arc across forewing. Male underside pale orange with brighter areas towards margin and hindwing 'tail', and across middle of hindwing. This brighter area is edged with white 'hairstreaks', one of which is continued on forewing. Underside pattern in female very similar, but ground colour is much brighter. **DISTRIBUTION** Occurs in most of Europe, but absent from the far south and much of Scandinavia, except the south of Norway, Finland and Sweden. In Britain occurs mainly in central-southern and south-western England and south-west Wales. In Ireland occurs close to the west coast.

HABITAT AND HABITS Found along wide woodland rides, woodland edges and hedgerows where blackthorn grows. Most populations are very thinly spread so it is quite rare to see more than one or two together. Main emergence period is mid- to late July and lasts through August into September. Males congregate high on 'master trees', where they attract females and mate with them. These then disperse to lower levels, flying hesitantly along blackthorn hedges and stopping occasionally to 'test' a twig and lay an egg. Both sexes feed on honeydew on leaves of ash or oak, but can also be seen taking nectar from flowers, especially bramble. The chrysalis is attended by ants, but many are eaten by small mammals.

Female

Male

Underside

Purple Hairstreak ■ *Favonius quercus*

DESCRIPTION Upperside brown with (in female) two wide swathes of purple-blue scales on forewing, while purple scales on male are more widely distributed over wing surface, but only visible at certain angles. Underside silver-grey with irregular white line across both fore- and hindwing, and with black-centred orange spot close to small 'tail' at rear. **DISTRIBUTION** Common and widely distributed in Europe, extending as far north as southern Sweden, Norway and Finland. In Britain found in Wales, and central and southern England, becoming more localized into Scotland. Scarce and patchily distributed in Ireland. **HABITAT AND HABITS** Characteristic habitat is oak forest, but also found on scrub oaks in uncultivated corners, hedgerow oaks and isolated mature oaks in parkland. Butterflies spend much of the day settled on leaves in the oak canopy (or, often,

on adjacent ash trees), where they feed from honeydew, occasionally making short flights. By late afternoon they become more active, with males occupying perches from which they dart out to intercept passing females. The latter spend more time basking with open wings between episodes of egg-laying on oak twigs. This is the best time of day to gain close views of them, especially in areas with scrub oak where they are likely to come closer to the ground. In some years they leave the trees to take nectar from nearby flowers, especially bramble. The chrysalis is said to be tended by ants in their nest, and the adult butterflies emerge from early July, reaching a peak towards the end of that month.

Female

Male

Female showing underside

Ilex Hairstreak ■ *Satyrium ilicis*

DESCRIPTION Upperside very dark brown, often with faint orange tint on female forewing border. Underside brown with row of black-bordered orange lunules close to hindwing margin. Irregular white 'hairstreak' on hindwing underside, and fine white line along

hindwing margin. Hindwings each bear short 'tail'. **DISTRIBUTION** Common and widely distributed in most of mainland Europe, but becomes more localized northwards to Denmark and the Baltic states, with colonies on the southern tip of Sweden. Absent from Britain and Ireland. **HABITAT AND HABITS** Typical habitat is deciduous woodland, where it occurs in open glades and sunny edges. Butterflies emerge from late June onwards, reaching a peak in mid-July. They spend much of their time at rest in the tree canopy, always with wings closed, but can often be seen taking nectar from flowers of bramble, thistles or knapweed. The eggs are laid on twigs of oak, and the caterpillars feed on buds and leaves.

Black Hairstreak ■ *Satyrium pruni*

DESCRIPTION Upperside in both sexes dark brown, with orange lunules towards wing margins of both fore- and hindwings in females, but restricted to hindwings in males. Underside quite similar to that of the White-letter Hairstreak (see p. 63), but the white 'hairstreak' forms a less convincing 'W' shape, and there is a row of black spots along the inner edge of the more extensive orange band on the hindwing. **DISTRIBUTION** Widely distributed in Europe except for the far south and north. In Scandinavia found in southern Finland and Sweden. In Britain natural populations are restricted to a swathe of countryside in the East Midlands. Absent from Ireland. **HABITAT AND HABITS** Found in large stands of blackthorn in open, sunny but sheltered situations in woodland or on woodland edges. The restricted distribution in Britain is a legacy of traditional woodland management in

this area, though many colonies have been lost because of more intensive management or abandonment. Woodland management designed to support the species is now showing success. Flight period is relatively brief, from the second week in June to early July, with some individuals appearing quite worn even by the third week in June. Butterflies spend much of their time in mature trees, feeding on honeydew, but can also usually be seen taking nectar from wild privet, dog rose and bramble. Like White-letter Hairstreak butterflies, they always settle with wings closed. Females lay their eggs on twigs of blackthorn. Caterpillars hatch in March or April and feed on flower buds, graduating to leaves as they develop.

Blue-spot Hairstreak ■ *Satyrium spini*

DESCRIPTION Upperside in both sexes dark brown with small orange spots close to rear margins of hindwings. Female has an obscure and variable orange area on each forewing. Underside pale brown or greyish-brown, with well-marked white streak across both wings. Small orange lunules along margin of hindwing, and large blue spot at rear. Pair of fine 'tails' projects from rear margin of hindwings. **DISTRIBUTION** Widespread in southern and central Europe, but absent from the north-western fringe, including Britain and Ireland, and from Scandinavia. Northern limits are reached in Luxemburg and Belgium (where it may now be extinct) and, in the north-east, Latvia, where it is rare. **HABITAT AND HABITS** Inhabits scrubby fields, track sides and woodland edges with abundant flowers. Like most other hairstreaks, almost always settles with its wings closed, so identification relies on the distinctive blue spot referred to in its common name. The adults can often be seen taking nectar from flowers such as bramble in May–late July, according to latitude and altitude. Caterpillars feed on buckthorns (*Rhamnus* species).

White-letter Hairstreak ■ *Satyrium w-album*

DESCRIPTION Upperside dark brown in both sexes, with small pale area of scent scales on male forewing. Underside pale brown, a little darker in male than in female. Cutting across both fore- and hindwing is a white line that forms the letter 'W' on the hindwing. Row of orange lunules towards rear of hindwing margin, which subtends two or three black marginal spots, one of which usually has a dusting of bright blue scales. Each hindwing bears small 'tail' at rear. **DISTRIBUTION** Occurs widely in Europe, though local in southern Europe, and absent from southern Spain. Northern limit is southern Scandinavia. In Britain locally distributed in south-east and central England, and East Anglia northwards to Yorkshire, and has a scattered distribution in Wales. Absent from Ireland. **HABITAT AND HABITS** Butterfly of woodland rides and edges as well as hedgerows with elm trees. Where associated with mature elms, the butterflies spend much of their time in the canopy, where they feed on honeydew and are rarely seen except when, as happens in some years, they descend to take nectar from creeping thistle or bramble. Females lay their eggs on or close to flower buds of elm, preferring wych elm, but also using other elm species. Since the 1970s the depredations of Dutch elm disease have greatly affected this butterfly, though wych elm seems to have been more resistant. Fortunately, if dead elms are left in place they usually regrow from suckers, and these can be used by White-letter Hairstreaks once they become sufficiently mature to flower. However, the saplings then succumb to the disease, so the butterfly can maintain its population only if there are substantial stands of the tree in different stages of regrowth. Flight period from early July to early August. It can be located by looking for its brown triangular outline on leaves or flowers close to elm sucker growth.

Purple-shot Copper ■ *Lycaena alciphron*

DESCRIPTION This butterfly has several distinct forms in different parts of Europe, but male of typical form in northern part of its range has orange uppersides with faint black

Male showing underside

Male

spots, the whole suffused with pale violet-blue scales. Female upperside dark brown with row of orange lunules around margin of hindwings. Underside hindwing grey with row of orange spots (sometimes fused to form narrow band) around outer margin, with adjacent black spots. Further in there is a row of white-ringed black spots, and another cluster close to wing base. **DISTRIBUTION** Widespread in southern, central and eastern Europe, but absent from or very rare in north-west. Reaches northern limit in Europe in the Baltic states, where it is very localized. Absent from Britain, Ireland and Scandinavia. **HABITAT AND HABITS** Inhabits flower-rich meadows, sheltered hollows in hillsides and, sometimes, open woodland clearings and roadside verges. Butterflies are active, taking nectar from meadow flowers such as ox-eye daisy, and fly in one annual generation from June to end of July. Caterpillars feed on leaves of sorrel *Rumex acetosa*, as well as other plants in the dock family, and have strong associations with ants.

Large Copper ■ *Lycaena dispar*

DESCRIPTION Male upperside fiery copper-red, with narrow black border and small black spot in centre of each forewing. Female has slightly less flamboyant orange ground colour to forewings, with black spots and black border. Hindwings black with orange border in most of Europe, but British and Dutch forms have more extensive orange flush on hindwing. Underside hindwing grey with black-bordered set of orange spots around margin – much more prominent in British and Dutch forms. **DISTRIBUTION** This iconic species was once widespread across Europe in several distinct subspecies. It is still widespread but increasingly scarce in central and eastern Europe, north to Finland. In western Europe it persists in scattered populations in France and the Netherlands. The British form (ssp. *dispar*) was once fairly widespread in the wetlands of East Anglia, but has been extinct since 1864. **HABITAT AND HABITS** Inhabits various types of wetland, including fens and marshes, wet meadows, and borders of lakes and rivers. In the fens of eastern England the butterfly was associated with traditional practices of peat and reed-cutting, but it became extinct in the middle of the 19th century as these practices died out and the fens were drained in favour of arable agriculture. Attempts to introduce the similar Dutch subspecies (*batava*) to suitable habitat in England were successful for quite long periods, while conservation management in the Netherlands seems to have at least halted the decline of the butterfly there. Caterpillars in Britain fed exclusively on water dock, but elsewhere they use several other robust dock species. In northern Europe the flight period is late June or early July to middle of August.

Male

Underside

Female, f. batava

Violet Copper ▪ *Lycaena helle*

DESCRIPTION Male upperside dark, with orange-and-black markings on forewing greatly obscured by dark scaling, but orange band to rear of hindwing is usually well marked. Wing

Underside

Female

surfaces suffused with violet, visible only at certain angles. Female has characteristic 'copper' pattern to upperside, orange on forewing with black spots, and dark brown-black hindwings with orange band. Orange on forewing suffused with dark scales, especially towards wing base, and variable amounts of violet spots and dusting are superimposed. Underside forewing orange with black spots, hindwing brownish with black spots and orange border. **DISTRIBUTION** This rare and beautiful species has only scattered populations in western Europe, but is more widespread in central and eastern Europe, extending northwards to central Sweden, Norway and Finland. Absent from Britain and Ireland. **HABITAT AND HABITS** Inhabits humid meadows, moist glades and open banks along streams and bogs in woodland, where the food plants of the caterpillars grow in abundance. There can be few more awesome spectacles for the butterfly lover than a sunny glade populated by numbers of this species. They settle, often with wings open, on leaves of bistort, or take nectar from its flowers. Butterflies emerge in May and can be seen until middle of June. Females lay their eggs on leaves of bistort *Polygonum bistorta*, or alpine bistort *P. viviparum* in the north of their range. This species is seriously threatened throughout Europe by drainage and changes in forestry management.

Purple-edged Copper ■ *Lycaena hippothoe*

DESCRIPTION Males have bright orange uppersides, with central black spot in forewing and purple tints along leading edge of forewing and among darker scales on hindwing. In northern Europe females are brown with vague areas of dull orange towards the middle of the forewings, and rows of faint dark spots. Orange band along rear margin of each hindwing. Underside hindwing grey with black spots and orange marginal band; forewing orange with black spots. **DISTRIBUTION** Widespread in suitable habitats throughout most of Europe, including much of northern Europe, with the exception of Britain, Ireland and north-western France. **HABITAT AND HABITS** Favours damp, flowery meadows where the food plants of its caterpillars (sorrel and various docks) grow. Like many meadow species it is threatened by spread of intensive agriculture, especially in lowlands. Flight period is from middle of June to early August.

Male

Underside

Female

Small Copper ■ *Lycaena phlaeas*

DESCRIPTION The shining copper-coloured upperside forewing first catches the eye. This has a dark border and an array of black spots. Hindwing mainly black, but with orange band at rear. In some individuals there is a row of bright blue spots aligned with the orange band. Underside more discreetly coloured, with pale orange-brown hindwing scattered with faint darker spots. Underside forewing brighter orange, with black spots mimicking upperside. **DISTRIBUTION** One of the most familiar and widespread of all butterflies that occur in Europe, curiously absent from large parts of central Sweden and Norway, but quite widespread further north. In Britain and Ireland found everywhere except in uplands of northern England and Scotland, and some of the islands. **HABITAT AND**

HABITS Habitat is dry grassland and heaths with well-drained soils. Also found on coastal dunes, worked-out sand quarries, roadside verges and urban brownfield sites, usually sparsely vegetated and with patches of bare ground. Males perch prominently and actively challenge passing insects, including females of their own species, which they pursue with a very rapid flight. Females are less active, and lay their eggs on food plants of their caterpillars, either common sorrel or sheep's sorrel *Rumex acetosa* and *R. acetosella*, depending on availability. Butterflies first appear in May (or later in the north), and produce a more numerous second brood by mid-summer, usually with a subsequent third brood in autumn.

Form with blue spots

Underside

Sooty Copper ■ *Lycaena tityrus*

DESCRIPTION Quite variable, tending to be darker further north and at high altitudes. Male upperside dark brown with indistinct black spots, and a variable row of orange lunules subtending black spots around border of hindwing. Female similar to male, but usually with a more prominent orange border, often extending to forewings. In some forms the female has a darkened orange ground colour on the forewing. Underside ground colour fawn or slightly yellowish-brown with black spots, and black-bordered row of orange lunules along wing borders. Wing shape distinctive, with a rather pointed tip to hindwing and relatively narrow forewings, especially in male. **DISTRIBUTION** Widespread and common in most of Europe, reaching north to Denmark and the Baltic states, and apparently now extending its range northwards into southern Finland. Absent from the rest of Scandinavia, Britain and Ireland. **HABITAT AND HABITS** Found in dry meadows, open glades in woodland, scrubland and other flowery places where the caterpillar's food plant grows. There are just two flight periods in northern Europe, from mid-May to mid-June and again from mid-July to early September. The eggs are laid on sorrel *Rumex acetosa* and probably other plants in the dock family.

Male

Underside

Female

Scarce Copper ■ *Lycaena virgaureae*

DESCRIPTION Male upperside brilliant fiery copper-red, quite similar to that of the Large Copper (see p. 65), but in most parts of Europe there is no central black spot, and black border widens steadily to tip of forewing. Female more subtly coloured orange than male,

with black spots on forewing, row of spots in darkened forewing border (plain black in the Large Copper), and orange band on hindwing. Underside very distinctive: orange-brown on hindwing, with scattered small black spots and row of white spots towards middle of wing.

DISTRIBUTION A fine butterfly found in most of mainland Europe, but absent from southern Spain, north-western France, Britain and Ireland. Further to the east it occurs in Denmark, the Baltic states and Scandinavia, except the far north.

HABITAT AND HABITS Found in flower-rich grassland such as unfertilized meadows, but also on wide roadside verges, woodland edges and open glades. Flight period is July–August, and the caterpillars feed on sorrel and sheep's sorrel.

Male

Underside

Female

Long-tailed Blue ■ *Lampides boeticus*

DESCRIPTION Male upperside shimmering violet-blue with narrow black borders. That of female grey-brown with variable blueish flush in basal area of wings. Both sexes have fine 'tails' leading from rear margins of hindwings, subtended by small black spots. Underside very distinctive, with small, fawn-brown blocks of colour, edged with narrow white cross-markings, and prominent white band. Black spots close to tails on underside usually have a few silver scales. **DISTRIBUTION** Characteristic species of southern Europe, probably fully resident only in the Mediterranean region. However, it is strongly migratory, and occasionally reaches southern Britain, where it may establish short-lived breeding populations. **HABITAT AND HABITS** In the south favours hot, dry, grassy areas with flowers, often on track sides or among patches of scrub, but further north can be seen in parks and gardens, along railway cuttings and roadside verges, and in other marginal habitats. In its Mediterranean haunts it is on the wing all year round and has several generations a year. Further north it is more usually seen from mid-summer onwards. Caterpillars feed on a wide range of plants in the pea family (Fabaceae).

Female

Mating pair

Small Blue ■ *Cupido minimus*

DESCRIPTION As its name suggests, this is one of our smallest butterflies. Male upperside plain grey-black, with dusting of violet blue scales, most concentrated around wing bases. Females dark brown and lack blue tints. Both sexes have fine white fringes to wings. Underside silver-grey with regular row of black spots towards outer border of forewing, and more irregular row in hindwing, with further spots towards wing base. Usually a flush of blue-green scales close to body. **DISTRIBUTION** Widespread but often very local butterfly across Europe northwards to southern Scandinavia, with outlying populations within the Arctic Circle. In Britain most populations are in southern and central England,

and south Wales, with scattered localities up to northern Scotland. In Ireland, too, it is limited to scattered colonies. **HABITAT AND HABITS** Limited to chalk and limestone grassland, often in small colonies in sheltered hollows on south-facing downs, limestone pavement, old quarries, railway cuttings, coastal cliffs and dunes. Butterflies emerge from mid-May, reaching a peak in early to mid-June, sometimes with a second generation in summer. Males occupy perches on tall grasses or shrubs from which they fly up to intercept females. After mating, females spend much of their time on flower-heads of the food plant of their caterpillars – kidney vetch. They lay their eggs among the florets and also take nectar from the flowers, as well as from other downland flowers such as horseshoe vetch, bird's-foot trefoil and sainfoin.

Underside

Male

Female

Short-tailed Blue ■ *Cupido argiades*

DESCRIPTION Upperside of male blue with narrow black border and small black spots along edge of hindwing. Female dark brown with dusting of blue scales across basal area of wings. Inconspicuous short 'tail' to each hindwing. Underside pale grey-blue, sometimes with brownish suffusion, and with array of black spots on fore- and hindwings.

A distinctive feature is two or three bright orange spots towards the hindwing margin, each subtending a black marginal spot. **DISTRIBUTION** Occurs in most of mainland Europe, though absent from the southern Iberian peninsula. Northwards reaches as far as the Baltic states and southern Finland, but absent from Norway, Sweden, Denmark, Britain and Ireland, except as occasional immigrants and released captives. **HABITAT AND HABITS** Inhabits warm, open slopes, fallow fields, flowery meadows and woodland edges, mainly at low altitudes. Flight period is from early May to late June, then in July and August. Females lay their eggs on bird's-foot trefoils, red clover and other plants in the pea family.

Underside

Male

Holly Blue ■ *Celastrina argiolus*

DESCRIPTION Male upperside shining sky blue, with narrow black borders to forewings. Female ground colour also blue, but she has wide black borders on forewings, and row of black spots to rear of hindwings. Underside pale grey-blue with row of small black spots on forewing, and more irregular scattering of spots on hindwing – rather similar to underside pattern of the Small Blue (see p. 72). **DISTRIBUTION** Common and familiar butterfly throughout Europe, more localized in the north and west of Scandinavia. In Britain generally common in England and Wales, reaching up to southern Scotland. Rather patchy distribution in Ireland. **HABITAT AND HABITS** Roams across both urban and rural landscapes, and found in parks and gardens, woodland edges and tall hedgerows, wherever the food plants of its caterpillars occur. Males are relentless patrollers of a sunny hedgerow or woodland edge, in search of females, stopping occasionally to bask in the sun

or take nectar from a flower. Females also bask with wings partially open, especially when the sun re-emerges after a spell of cloudy weather. There are two generations each year (sometimes three), from mid-April to June, then from late July through August. The eggs are commonly laid on the flower buds of holly in spring, and ivy in autumn, but a wide range of shrubs is also used, including dogwood, spindle, snowberry and *Pyracantha*. The species experiences great population explosions followed by crashes across most of its geographical range. These appear to be caused by a specialist wasp parasite whose population alternates in abundance with that of the Holly Blue.

Underside

Male

Female

Large Blue ■ *Phengaris (Maculinea) arion*

DESCRIPTION Among the largest of the European blues. Upperside blue with wide black border, an arc of tear-shaped black marks on outer part of forewing, and black central spot.
These markings are usually more extensive in females than in males. Underside grey with double row of small black markings following wing borders, and array of larger black spots on both fore- and hindwings. Also an iridescent green-blue flush on basal areas of hindwings. **DISTRIBUTION** Although still quite widespread across much of Europe, including southern Scandinavia, this iconic species is in decline everywhere. In Britain it was once abundant in some localities, especially in the south-west, but became extinct in 1979. It has since been reintroduced. **HABITAT AND HABITS** In Britain the butterfly inhabits south-facing dry hillsides, with short turf, growths of thyme and strong populations of a red ant, *Myrmica sabuleti*. Elsewhere in Europe it can be found in a wider range of habitats, such as flowery woodland rides and even mountain slopes. Butterflies are on the wing from early June to mid-July, the males spending much of the morning patrolling the hillside, flying low in search of freshly emerged females. Females lay their eggs singly among the flower buds of wild thyme, and the resulting caterpillar feeds on this plant. Later it drops to the ground, where it is found by ants attracted by its sugary secretions and ant-like scent. The caterpillar is taken into the ant's nest, where it completes its development by feeding on ant larvae. It goes on to form a chrysalis that 'sings' to the ants, and is surrounded by them when it finally emerges as an adult. The full details of this complex life story, including the need for the right species of ant to discover the caterpillar, were discovered by leading lepidopterist Jeremy Thomas, just too late to save the butterfly from extinction in Britain, but of value in informing the successful reintroduction of the species.

Female

Male

Underside

Male

Female, underside

Alcon Blue ■ *Phengaris (Maculinea) alcon*

DESCRIPTION Male upperside blue with narrow black border, while female's is dark brown, usually with sparse dusting of blue scales on basal area of wings. Underside pale brown with faint double row of darker spots along wing margins, and pattern of dark spots similar to that on underside of the Large Blue (see p. 75), though usually smaller. Basal blue flush on underside less extensive than on the Large Blue. **DISTRIBUTION** Widespread but very local in Europe, and absent from the far south. Occurs in Denmark and southern Sweden, but absent from Britain and Ireland. **HABITAT AND HABITS** Habitat includes wet meadows, marshes and lake margins where marsh gentian *Gentiana pneumonanthe* grows. However, the Alcon Blue belongs to a species complex, with some close relatives occupying dry hillsides where other gentian species are used. In the north it can be seen from late June to early July. The eggs are laid singly, but often with many scattered over the leaves and flowers. The resulting caterpillars feed on the tissues of the gentian before dropping to the ground. Like the caterpillars of the Large Blue, they are taken into the nests of ants where they complete their development. It seems to be less dependent on being discovered by a single species of ant, several species of *Myrmica* being listed as hosts.

Dusky Large Blue ■ *Phenargis (Maculinea) nausithous*

DESCRIPTION Male upperside dusky purple-blue with row of black spots on each wing, and wide grey-black borders. Female upperside uniform dark brown. Underside plain brown with single row of small, white-ringed black spots on both fore- and hindwings.
DISTRIBUTION This species is very rare and vulnerable in central and eastern Europe, with outlying populations in Bulgaria and Spain. It is at the north-western edge of its

Female, egg laying

range in the Netherlands, and is absent from Scandinavia, Britain and Ireland. **HABITAT AND HABITS** The specialized habitat of this butterfly is wet meadows and marshes, often close to lakes. Adult butterflies are exceptionally inactive, spending much of their time with wings closed on the flower-heads of the food plant of their caterpillars. This is greater burnet *Sanguisorba officinalis*. The eggs are laid on the flower-heads, and the caterpillars initially feed on this plant, later being taken into the nests of a red ant, *Myrmica rubra*. Already rare because of its habitat requirements, this species is increasingly threatened by drainage of its few remaining localities, but now has strong conservation protection. Flight period is late June–August.

Scarce Large Blue ■ *Phenargis (Maculinea) telejus*

DESCRIPTION Male upperside a beautiful, shimmering pale silver-blue, with tear-shaped black spots and narrow black border to wing. Female upperside dark blue with wide borders that obscure blue colouration in places. Underside grey to grey-brown, darker in females, with two rows of small, white-ringed black spots on outer part of both fore- and hindwing, and several other small spots closer to hindwing base.**DISTRIBUTION** The Scarce Large Blue justifies its English name, being rare and localized in central and eastern Europe. In the north it has outlying populations in Latvia and Lithuania, and is at its northern limit in Belgium and the Netherlands, where it is rare or extinct. Its habitat overlaps closely with that of the Dusky Large Blue (see p. 76),

and both species frequently occur together.
HABITAT AND HABITS Like Dusky Large Blue butterflies, those of this species are very closely associated with the flower-heads of greater burnet, and settle with their wings closed. Although their demanding habitat requirements make them a very localized species, they can be quite abundant where they do still occur. To see the splendid upperside it is necessary to locate roosting specimens and watch them early in the day as the first rays of the sun reach them, and their wings slowly open. Caterpillars feed on burnet plants in their early stages and later are taken into ants' nests. Flight period is from mid-June through July into August.

Female, egg laying

Male

Female

Green-underside Blue ■ *Glaucopsyche alexis*

DESCRIPTION Male upperside plain blue with black margin that widens towards apex of forewings. Female upperside brown, usually with pale blue scales over at least the basal areas of both fore- and hindwings. Underside plain grey with extensive flush of turquoise scales on basal area of hindwing, and one row of white-ringed black spots on forewing that

Underside

increase in size towards rear edge of wing. This is matched by row of much smaller black spots on hindwing (which may be reduced in number or absent). **DISTRIBUTION** In Europe this species has a discontinuous distribution, being widespread in southern and central Europe, but either absent or very localized further north up to southern Sweden, Norway and Finland, where it is again quite widespread. It is absent from Britain and Ireland. **HABITAT AND HABITS** Inhabits a range of flowery habitats from roadside verges to small meadows and light scrub, where the food plants of its caterpillars (various vetches and other legumes) abound. In northern Europe the butterfly is on the wing from mid-May to early July.

Male

Eastern Baton Blue ■ *Pseudophylotes vicrama*

DESCRIPTION Upperside of male bright silvery blue with slightly darkened borders and black spots close to margins of hindwings. Short, linear black marks at centre of both fore- and hindwings. White fringes are part-laddered with black, especially on forewings. Underside silver-grey with white-ringed black spots and row of red spots along hindwing margin. This is edged with black on both sides. **DISTRIBUTION** Predominantly a species of southern and eastern Europe, with outlying scattered populations in the Baltic states and a few inland localities in Finland. In western Europe its place is taken by its sister species, the Baton Blue *S. baton*, which is indistinguishable in general appearance. Neither species occurs in Britain or Ireland. **HABITAT AND HABITS** Inhabits dry grassland, glades in light woodland, rocky slopes and sandy heaths, usually with wild thymes. The butterfly has just one generation in a year, in June–July, in the northern part of its range, but it has two broods, in April and August, further south. Caterpillars feed on several species of thyme and also calamints but may not be able to complete their life-cycle without their strong association with ants.

Underside

Male

Female

Chequered Blue ■ *Scolitantides orion*

DESCRIPTION Upperside of male black with variable suffusion of blue scales over most of wing surface. Row of black spots bordering both fore- and hindwing, each usually with halo of pale blue or whitish scales. Female similar to male, but with less blue scaling. Wing fringes pure white with prominent dark laddering. Underside grey-white, with large, strongly contrasting black spots and bright orange band close to border of hindwing.
DISTRIBUTION A rather local species with a discontinuous distribution in Europe. Widely distributed across southern and central Europe. Occurs again in southern Finland, with a few localities in southern Sweden and Norway, but absent from intermediate latitudes, and from Britain and Ireland. **HABITAT AND HABITS** Inhabits dry, sparsely vegetated, rocky slopes, quarries and gullies, often coastal in the northern part of its range. Flies close to the ground in exposed habitats, basking on lichens or rock surfaces. Caterpillars feed on orpine and other sedums, and adults are on the wing in May–June, sometimes with a small second brood in summer.

Underside

Male

Arctic Blue ■ *Plebejus aquilo*

DESCRIPTION Male upperside blue-grey with small black central spot in forewing and vaguely defined paler markings towards wing margins. Female similar to male but pale grey-brown in colour. Underside slate-grey with vestigial black markings, especially on forewing, and often one or two orange marginal marks on hindwing. On hindwing and sometimes forewing, black spots typical of the *Plebejus* group are replaced by white patches.

DISTRIBUTION Very localized species, in Europe found only in the far north of Sweden, Norway and Finland. However, some authors consider it to be a subspecies of the Glandon Blue *P. glandon*, which flies at high altitudes in the Alps, and it also seems to be closely related to similar butterflies that occur in the Pyrenees and mountains of southern Spain. **HABITAT AND HABITS** Rare and local on rocky mountain slopes with loose scree, as well as industrial spoil heaps in at least one site. Butterflies take nectar from bladder campion and stonecrop, and females lay their eggs on the leaves of purple saxifrage *Saxifraga oppositifolia*. Males settle on rocks with wings half-open, flying up to see off passing males, but also butterflies of other species. Flight period is usually from late June to early August.

Male

Female

Underside

Silver-studded Blue ■ *Plebejus argus*

DESCRIPTION Very variable across different localities and habitats, but male is blue on upperside, usually with wide dark wing borders and row of black spots linked to border on hindwing. Tiny spur at apex of tibia in males, absent in the Idas Blue (see p. 84). Female is similar to that of the Idas Blue, but ground colour of underside is usually darker than in that species. Row of green-blue spots close to margin of hindwing, and array of relatively large black spots. **DISTRIBUTION** Very widespread and usually common throughout Europe, except northern Scandinavia, Ireland and large parts of Britain. There it is found mainly in southern and south-eastern England, with isolated populations in the south-west

and north of Wales. **HABITAT AND HABITS** Wide range of habitats, including calcareous grassland and wet or dry heathland, coastal cliffs and dunes. Flight period is from late June, reaching a peak in July with some individuals surviving through August, but in the south of its range it usually has two broods each year, in spring and high summer. Colonies are quite sedentary, and it is slow to occupy new territory. On heathland the caterpillars feed on gorse, ling and other heathers, and from the beginning are attended by black ants (genus *Lasius*), which take them into their nests, and appear to guard them when the adults first emerge. In Britain many limestone grassland colonies have been lost, but a distinct subspecies survives in north Wales and is attended by a different species of *Lasius*.

Male

Female

Underside

Reverdin's Blue ■ *Plebejus argyrognomon*

DESCRIPTION Very similar to the Idas Blue (see p. 84), with narrow black borders to blue uppersides in male, and blue-green marginal spots on underside hindwing. In areas where the two species overlap, the slightly larger size, paler, more uniform blueish-white underside of the male, relatively small black spots, and orange spots forming a complete series on the forewing are useful guides. **DISTRIBUTION** Fairly widespread across mainland Europe, but decidedly scarce in the north. Occurs in Latvia and Lithuania, and is very scarce in southern Sweden and Norway. Absent from most of the rest of northern Europe, including Britain and Ireland. **HABITAT AND HABITS** Inhabits a wide range of flowery grassland, including roadside verges and cuttings, open spaces in woodland and shrubby downland, on calacareous soils. In Scandinavia the caterpillars feed on wild liquorice *Astragalus glycyphyllos* and have just one flight period, from early July to middle of August. Elsewhere crown vetch is used, and in the southern part of its range there are two generations each year. The caterpillars are attended by several ant species.

Underside

Male

Female

Idas Blue ■ *Plebejus idas*

DESCRIPTION One of three similar species with blue uppersides in males, and a row of metallic silver-blue or green spots close to margin of underside hindwing. Black border to upperside in male. Female looks very similar to female of the Silver-studded Blue (see p. 82), being brown with variable amounts of blue scaling, and with dull orange markings near hindwing margins. In the north of range, subspecies *lapponica* is paler blue in male, but underside ground colour is slightly darker, giving a strong contrast with the white-and-orange bands. Female upperside has very few blue scales. **DISTRIBUTION** Widespread across mainland Europe, including Scandinavia and most of northern Europe, but absent from Britain and Ireland. **HABITAT AND HABITS** Inhabits flowery meadows, hillsides and heaths, wherever there are suitable food plants for the caterpillars, and appropriate ant species. The life-cycle takes two years in the north, the caterpillars feeding on crowberry (*Empetrum* species), and adults fly from late June to late August. Further south there are often two broods in a year, and various plants are used by the caterpillars, including bird's-foot trefoil, broom, kidney vetch and heathers. As is the case with many species in this family, the caterpillars are attended by ants.

Male, lapponica

Female

Underside

Cranberry Blue ■ *Plebejus optilete*

DESCRIPTION Males have violet-blue uppersides, with narrow black margin and clear white fringes. Females brown with variable dusting of violet-blue scales, especially towards wing bases. Underside of both sexes grey with black spots. Towards the rear of the hindwing margin is a distinctive orange patch subtending a turquoise-centred marginal black spot. **DISTRIBUTION** Widely distributed in the Alps, across central and northern Europe, including Scandinavia, but absent from Britain and Ireland. **HABITAT AND HABITS** Inhabits both boggy areas and dryer, flowery habitats, and on the wing from late June to as late as August, depending on latitude and altitude. In the northern part of its range it is active even in dull weather. Caterpillars feed on leaves of cranberry and other *Vaccinium* species, as well as marsh andromeda.

Female

Underside

Male

Alpine Blue ■ *Plebejus orbitulus*

DESCRIPTION Male upperside brilliant clear blue, with narrow black wing margins and pure white fringes. Female dark brown, sometimes with basal blue flush. Male underside grey-blue with small black spots on forewing, and two rows of very distinctive white oval spots on hindwing. Female underside similar but with brown ground colour.

Female, underside

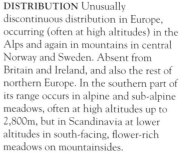

DISTRIBUTION Unusually discontinuous distribution in Europe, occurring (often at high altitudes) in the Alps and again in mountains in central Norway and Sweden. Absent from Britain and Ireland, and also the rest of northern Europe. In the southern part of its range occurs in alpine and sub-alpine meadows, often at high altitudes up to 2,800m, but in Scandinavia at lower altitudes in south-facing, flower-rich meadows on mountainsides.

HABITAT AND HABITS Males are often seen communally drinking from damp soil or shingle, while females search out the milk-vetches (*Astragalus* species) on which they lay their eggs. Flight period is normally July, but can be earlier or later according to weather and altitude.

Male

Male showing underside

Brown Argus ■ *Aricia agestis*

DESCRIPTION Upperside of both sexes brown with small black spot in centre of each forewing, and row of orange lunules close to wing margins. These are more prominent in females and often much reduced, especially on forewings in males. Underside grey-brown (brown in females), with black spots and row of orange lunules along margins of both fore- and hindwings. No black spot between central spot ('discal' mark) and wing base on forewing underside; two black spots close to leading edge of hindwing roughly at right angles to edge. These features distinguish the Brown Argus from brown forms of the female Common Blue (see p. 96). **DISTRIBUTION** This is a complex species, occurring in several named forms throughout Europe northwards to Denmark, the southern tip of Norway and Sweden, and southern Britain. Absent from Ireland. **HABITAT AND HABITS** Found on rough grassland habitats such as roadside verges and railway cuttings, flowery meadows, grazing marshes and heathland on acid or calcareous soils. On chalk or limestone grassland the favoured food plant of the caterpillars is common rock-rose *Helianthemum nummularium*, but they also use various species of cranesbill (*Geranium* species) and storksbill (*Erodium* species). There are usually two generations each year, with flight periods from late May to late June, and again from beginning of August into September. In Britain the species suffered a sharp decline from the 1950s until the '80s, but regained much lost ground from the early '90s onwards. It is now widely distributed up to north Yorkshire.

Underside

Male

Female

Northern Brown Argus ■ *Aricia artaxerxes*

DESCRIPTION Very variable according to locality. Upperside brown, often with orange lunules close to wing borders, as in Brown Argus (see p. 87). However, in some populations lunules are reduced and may even be absent, especially on forewings. This form is frequent in Sweden. Underside pattern similar to that of the Brown Argus, but white-ringed black spots are often reduced in size (frequent in northern England) or absent altogether, just leaving an array of white spaces (frequent in Scotland). In Scotland both males and females have clear white central spots on forewings (and also sometimes on hindwings). **DISTRIBUTION** Discontinuous European distribution, being widespread, often in mountains, in southern and central Europe, and also in the Baltic states and Scandinavia.

Scottish form

Absent from the far north of Scandinavia, but locally common in parts of northern England and Scotland. **HABITAT AND HABITS** Warm, sheltered hillsides with rocky outcrops, worked-out limestone quarries and lightly grazed grassland with shrubs. Butterflies bask with open wings in sunny conditions, generally making only short flights between flowers or, in the case of females, seeking out suitable plants of rock-rose (or, on some sites, cranesbills) on which to lay their eggs. There is one generation per year, flying (in northern Europe) from beginning of June to mid-August, peaking in late June or early July.

Male, Sweden

Underside, Scottish form

Geranium Argus ■ *Aricia eumedon*

DESCRIPTION Upperside dark brown in both sexes, often with faint orange lunules along rear edge of hindwing in females. Underside pale grey-brown with row of small black spots on both fore- and hindwings, and row of orange spots towards wing margin. Dusting of metallic green scales towards base of hindwing; white edging of central spot on underside hindwing extended to form white streak to outer edge of wing. In fresh specimens there is a fine white fringe around the wings. **DISTRIBUTION** Widespread throughout much of Europe, though absent from western and northern France, Belgium, the Netherlands, Denmark, Britain and Ireland. Generally distributed in Scandinavia, but becomes much more local or is absent in the far north. **HABITAT AND HABITS** Habitat is flower-rich meadows with cranesbills *Geranium sanguineum* and *G. sylvaticum*, with which the butterflies are closely associated, both as nectar sources and as the food plant of their caterpillars. Butterflies are on the wing from beginning of June to mid-July, depending on latitude.

Mating pair

Male

Female

Silvery Argus ■ *Aricia nicias*

DESCRIPTION Male upperside shining greenish-blue with wide dark borders and narrow bar at centre of forewing. Female upperside brown, usually with blue scales towards wing

bases. Underside pale grey-brown, with a few small black spots, a faint orange-brown row of lunules around hindwing margin, and a distinctive white streak. **DISTRIBUTION** European distribution sharply discontinuous. Occurs in the eastern Pyrenees, central Alps, and again in central Sweden and southern Finland. Absent from Britain, Ireland and the rest of northern Europe. **HABITAT AND HABITS** Rare and local in its Scandinavian range, flying on damp, flower-rich meadows, often on south-facing slopes with calcareous soils. Males bask with open wings, while females lay their eggs on wood cranesbill or meadow cranesbill. Flight period is usually from beginning of July to mid-August.

Underside

Male

Amanda's Blue ■ *Polyommatus amandus*

DESCRIPTION Males have pale blue/turquoise uppersides, with fine black outlining of wing veins towards outer edges of wings, where they fuse with the narrow black border. Females more variable, sometimes wholly brown on upperside, and sometimes with blue scales on extensive areas of both fore- and hindwings, and orange spots close to borders of hindwings. Male undersides distinctive, with row of very small black spots across both fore- and hindwings, and row of small, often obscure orange lunules along hindwing margin. Flush of turquoise scales around wing base. Females more boldly marked on underside than males. **DISTRIBUTION** Occurs widely in southern and central Europe, as well as Denmark, Germany, Poland and the Baltic states. In Scandinavia often common in the south and east of Sweden, south-eastern Norway and southern Finland. Does not occur in Belgium, the Netherlands, Ireland or Britain. **HABITAT AND HABITS** Inhabits open, flowery meadows, woodland clearings and, in Sweden, the margins of bogs, and flies from mid-June to late August. Takes nectar from meadow vetchling and other flowers in the Fabaceae family (vetches and clovers), and also lays its eggs on these plants.

Male

Blue female

Male, underside

Brown female

Adonis Blue ■ *Polyommatus bellargus*

DESCRIPTION Male glittering turquoise-blue on upperside, changing tint as it moves around a flower-head in the sun. This colour and the laddering of its outer wing-fringes mark it out from all other species. Female upperside warm brown with orange marginal lunules and a variable dusting of blue scales. Very similar to female of the Chalkhill Blue (see p. 93), but area between marginal spots and edge of hindwing is infilled in blue. Underside boldly marked in both sexes, and ground colour is pale brown – darker in female. **DISTRIBUTION** Widely distributed in Europe, becoming more local northwards to southern England, its north-western limit. Absent from Ireland, Scandinavia and Denmark. **HABITAT AND HABITS** Close to its climatic limit in northern Europe,

and confined to the hottest localities – usually sheltered, south-facing downland, with close-cropped, often broken turf. Males make a vivid sight as they fly close to the ground in search of newly emerged females, or stop to take nectar from flowers such as marjoram. In Britain the caterpillars feed only on horseshoe vetch, but crown vetch is also used elsewhere. The association with ants is particularly close. There are two generations each year, with the butterfly being on the wing in mid-May–June, and again in July–September. In England the species suffered major habitat loss in the 20th century and extinction loomed by the 1970s. However, urgent conservation measures, combined with a recovery in rabbit grazing (and possibly climate warming), have resulted in a strong recovery.

Underside

Male

Female

Chalkhill Blue ■ *Polyommatus coridon*

DESCRIPTION Male upperside spectacular silver-blue, with indistinct darkened border to forewing, and row of marginal black spots on hindwing. Male underside pale blue-grey with brownish tint on hindwing, and iridescent green-blue close to wing base. The underside pattern of black spots and a marginal row of orange lunules is present, but usually the spots are smaller, and the orange is fainter than in related species. Female upperside brown with variable blue scaling, especially towards wing bases, and with some orange lunules close to margin of hindwing. Female underside strongly marked, with a rich brown ground colour. Upperside colour of males is distinctive, while females can be distinguished from Common Blues (see p. 96) by the prominent black-and-white laddering of the wing fringes, and from the Adonis Blue by the white (not blue) infilling of the space between the marginal markings on the upperside hindwing and wing edge. **DISTRIBUTION** Widespread across mainland Europe, though absent from southern Spain. Northwards reaches its limit in Lithuania, northern Poland and southern England, where it has very scattered populations on chalk and limestone hills. **HABITAT AND**

HABITS In the northern part of its range restricted to warm, sheltered areas of chalk or limestone grassland, frequently on lightly grazed, south-facing downland hills, but also on verges, disused railway tracks and open areas on woodland edge. Butterflies take nectar from carline thistle, knapweed, scabious, marjoram and other downland flowers. Males actively search for females, while the latter are more sedentary, laying their eggs on or close to the food plant of the caterpillars, horseshoe vetch. Both caterpillars and chrysalids are attended by ants. Flight period is from mid-July to early September in one generation.

Male

Female

Female, underside

Damon Blue ■ *Polyommatus damon*

DESCRIPTION Male upperside pale blue, with veins outlined in grey-brown towards the dark grey-brown borders. Wing fringes are white. Female upperside plain brown

with white fringes. Underside grey-brown in males, darker in females. Single row of white-ringed black spots across both wings, those on underside forewing being larger than those on hindwing. Prominent white streak across underside hindwing in both sexes. **DISTRIBUTION** Very discontinuous European distribution, occurring in mountainous areas of Spain, Italy, the Balkans and central Europe. In the north there are just isolated colonies in Latvia. **HABITAT AND HABITS** Inhabits grassy slopes, open woodland and scrub at moderate altitudes in mountains. There is one generation each year, and the flight period is in July and August. Caterpillars feed on sainfoins (*Onobrychis* species).

Male

Underside

Female

Turquoise Blue ■ *Polyommatus dorylas*

DESCRIPTION Male upperside shining turquoise-blue, with narrow black border and clear white fringes. Female upperside plain dark brown, often with small orange lunules close to rear margin of hindwings and with brown suffusion of white fringes on forewings. In both sexes undersides are distinctive. Wavy row of white-ringed black spots on both wings, with a central spot, often a white mark. Small orange spots towards margin of both wings, but border between these and outer margin of each wing is white. **DISTRIBUTION** Mainly a butterfly of southern and central Europe, but has a few outposts in the north: north-eastern Poland, Lithuania and south-eastern Sweden (notably Gotland). **HABITAT AND HABITS** Habitat used in the north is south-facing and sheltered slopes, shrubby limestone pavement and other warm habitats where the food plants of the caterpillars grow. Butterfly is very dependent on warmth in its northern range, flying only in full sunshine, and often basking on bare rock. The eggs are laid on kidney vetch *Anthyllus vulneraria* and *Melilotus* species.

Male

Male showing underside

Female

Common Blue ■ *Polyommatus icarus*

DESCRIPTION Male upperside bright shining blue, with fine black marginal line. Female brown with variable amounts of blue scaling (often more extensive in the north), and row of orange lunules along wing margins. Underside markings (black spots and marginal orange lunules) similar in both sexes, but ground colour is pale brown in females, blue-grey in males, often with a basal flush of iridescent scales. Brown females can easily be mistaken for the Brown Argus (see p. 87), but the presence of one or more spots between the central spot and wing base on the underside forewing, as well as the arrangement of spots on the hindwing, serve to separate the two. **DISTRIBUTION** Very widespread and often common throughout Europe, with many named local forms that some authors treat as distinct species. **HABITAT AND HABITS** Inhabits very wide range of unfertilized grassland,

including lightly grazed pastures, roadside verges, cuttings, open glades and woodland edges, south-facing hillsides on downland, disused quarries, lowland heaths, parkland and even gardens. Butterflies bask with open wings early in the morning and before dusk, as well as when the sun shines after cloudy weather. They take nectar from bird's-foot trefoil, clovers, knapweeds, marjoram and other flowers, according to habitat. In Britain the eggs are usually laid on bird's-foot trefoil, but elsewhere numerous other plants in the pea family (Fabaceae) are used, including rest-harrows and lucerne in Scandinavia. In northern Scandinavia there is just one brood per year, in July or August, but further south there are two broods, from mid-May to late June, and from July to beginning of September (with a third brood in some years).

Male

Underside

Female

Mazarine Blue ■ *Polyommatus semiargus*

DESCRIPTION Upperside of males is violet-blue, with very fine black wing veins that become more apparent towards the ill-defined black borders. Females usually have plain brown uppersides, sometimes with blue scales towards wing bases. Undersides of both sexes pale grey-brown

with wavy line of small black spots running across both fore- and hindwings. Scattering of green-blue scales towards base of underside hindwing.
DISTRIBUTION Occurs and is often common throughout Europe, except northern and western Scandinavia. Formerly occurred in south-western and central England and in south Wales, but became extinct during the 1870s for reasons that are not fully understood **HABITAT AND HABITS** Habitat is flowery meadows and clearings in woodland, and butterfly can be seen from mid-June to end of July. Caterpillars feed on red clover, and in some places on kidney vetch.

Underside

Male

Female

ADMIRALS, TORTOISESHELLS, EMPERORS, FRITILLARIES & BROWNS (NYMPHALIDAE)

This huge and very diverse family of butterflies is divided into two subfamilies, the Nymphalinae and the Satyrinae. These used to be treated as two distinct families, and for convenience, here they are treated separately.

NYMPHALINAE

This subfamily includes two distinct groups of butterflies. One group consists of the medium to large, brightly coloured, strongly-flying species that are given such names in English as 'emperors', 'admirals' and 'tortoiseshells'. Most species are primarily inhabitants of woodland, and many lay their eggs on trees or shrubs, notably willows, but also poplars and elms. The larger species in this group, such as the White Admiral, Red Admiral and Peacock, have average wingspans of 5.5cm, the Purple Emperor and Poplar Admiral being a little larger. The Small Tortoiseshell and Comma are medium-sized butterflies, at 4–4.5cm, while the Map is smaller at 4cm.

The second group is the fritillaries. These have a characteristic upperside pattern of orange with networks of black lines, or a combination of lines and spots. Many species have one or more pale bands running across the underside hindwing, and several have an array of silver spots or washes. Most of the fritillaries fall into two size ranges. The largest include those in the genus *Argynnis*, with wingspans of 5.5–6cm, the Silver-washed Fritillary being the largest in northern Europe, at 6–6.5cm. The smaller fritillaries include species of *Boloria*, *Melitaea* and *Euphydryas*, with wingspans averaging 3.5–4cm. The Marbled Fritillary and Queen of Spain are intermediate in size.

White Admiral ■ *Limenitis camilla*

DESCRIPTION Upperside very dark brown/black with obscure darker markings, and broad white band across both fore- and hindwings. Smaller white spots close to apex of forewings. Underside spectacularly beautiful, with alternating bands of red and white, and area of silver-grey close to base of hindwing. **DISTRIBUTION** Widespread in Europe, but

absent from most of Scandinavia. However, it reaches as far north as Denmark, southern Sweden and the Baltic states, and seems to be extending its range northwards. In Britain quite widespread in southern and central England, and Wales, with some evidence of recovery of lost territory. **HABITAT AND HABITS** Inhabits rides and glades in mature woodland, as well as woodland edges and tall hedgerows. Although it needs open spaces in woodland, it does not depend on coppicing. On sunny days the males, especially, glide contouring shrubs and trees. Females lay their eggs on honeysuckle in shady parts of woods, and adults fly from mid-June to late July or early August.

Poplar Admiral ■ *Limenitis populi*

DESCRIPTION Large, spectacular and elusive butterfly. Upperside velvety dark grey to black, with white band across hindwing and a more discontinuous one on forewing. White band often reduced or obscured, and may be absent altogether. Obscure rows of blue-and-orange markings around wing borders, particularly on hindwing. Underside a mixture of brick-red and blue-grey with an echo of white markings of upperside. **DISTRIBUTION**

Widespread though local in central and eastern Europe, but absent from many parts of the south. Quite widespread in south-central Sweden and Finland, and southern Norway, and seems to be extending its range northwards. Absent from Britain and Ireland, and apparently extinct in Denmark. **HABITAT AND HABITS** Woodland butterfly often inhabiting quite dense plantations in some areas, and found along forest roads bordered with aspen. Adults spend much of their time in the canopy, but males descend to feed on carcasses (including slugs squashed by forestry vehicles), mud, dung or human sweat. They have a short flight period, from late June to mid-July in the north, and the caterpillars feed on the leaves of aspen.

Underside

Purple Emperor ■ *Apatura iris*

DESCRIPTION This butterfly's name and celebrity is due to the brilliant purple sheen covering the surface of the upperside, visible only as light strikes it from certain angles. Male upperside brown, with narrow band of white across both wings and cluster of white spots close to apex of forewing. Small, orange-ringed black spot at rear of hindwing. Female very similar to male, but usually larger and lacking purple tints. Underside patterned with red and blue-grey, with toothed white streak across hindwing. Small eye marking at rear of hindwing, and a larger one on forewing. **DISTRIBUTION** Like the White Admiral (see p. 98), this fabled quarry of past generations of collectors is widespread in Europe, except many areas in the Mediterranean region. Reaches as far north as the Baltic states, southern Finland, southern Sweden and Denmark, and seems to be extending its range northwards.

Underside

In Britain its stronghold is in the woods of central southern England, but it has colonized new areas in recent decades. Absent from Ireland. **HABITAT AND HABITS** Butterfly of mature woodland. Usually a single population is distributed across several nearby woods. Butterflies are large, powerful flyers, often seen soaring high along a ride or, in the case of females, fluttering around a large sallow bush to lay eggs. Males descend to feed from dung, carrion or mud along woodland rides, and can easily be approached. Later in the day they gather around groups of 'master trees', and mating takes place in the canopy. Flight period is late June to mid-August, later in the north.

Male

Female

Lesser Purple Emperor ■ *Apatura ilia*

DESCRIPTION There are two colour forms of this butterfly. One is similar to the Purple Emperor (see p. 100) in colouring, while the other (form *clytie*) is orange-brown in ground colour, with yellow-orange replacing the white bands. In both forms males have a purple sheen. On both surfaces of hindwing, white (or yellow) band lacks tooth-like projection found in the Purple Emperor. Also, there are orange-ringed black spots in both fore- and hindwing upperside in the Lesser Purple Emperor (hindwing only in the Purple Emperor).

DISTRIBUTION Widely distributed in central and eastern Europe, but absent from large parts of the Mediterranean region in the south, and currently absent from Denmark, Sweden and Norway in the north. However, in the north-east it reaches as far north as the Baltic states and southern Finland, where its range is now expanding. It is absent from Britain and Ireland. **HABITAT AND HABITS** Woodland butterfly, often flying in the same areas as the Purple Emperor and, like it, descending to forest tracks to 'mud-puddle', or feed from dung or carcasses (or human sweat). In northern Europe there is one generation that flies from early June to late July (later further north), and the caterpillars feed on aspen *Populus tremula*, or other poplars and willows.

Male

Underside

Map ■ *Araschnia levana*

DESCRIPTION This butterfly has two strikingly different colour forms, one in the spring generation, the other in summer. Upperside of spring form has orange ground colour with white-edged areas of black towards bases of wings, and black spots together with a few white patches further out towards the margins. Summer brood is black with prominent white band across hindwing, irregular pattern of white markings on forewing and obscure orange-red lunules close to wing margins. Spring and summer undersides more similar, with intricate pattern of reds and browns crossed by white veins to give a map-like appearance. The rather angular wing-shape is distinctive. **DISTRIBUTION** Widespread and often common in central and eastern Europe, but absent from large parts of the south.

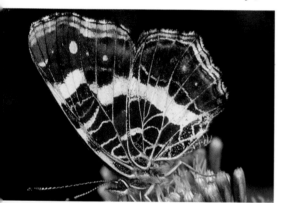

In northern Europe extends to Belgium, the Netherlands, Denmark, southern Sweden, south-eastern Finland and the Baltic states. Absent from Britain and Ireland. **HABITAT AND HABITS** Inhabits sheltered glades, clearings and rides in woodland, as well as scrubby pastures. Two generations each year in most places, in May–June, then in August–September. The seasonal difference in the appearance of the adults is determined by day length during the development of the caterpillars. The eggs are laid on stinging nettles.

Underside, summer

Summer

Spring

Red Admiral ■ *Vanessa atalanta*

DESCRIPTION Upperside black with brilliant red band across forewing, and orange-red marginal band around hindwing. Cluster of white spots close to apex of forewing, and some blue scaling towards outer wing margin. Black spots in each cell of hindwing marginal band, and blue spot at rear of each hindwing. Underside hindwing has intricate pattern of browns and greys, while forewing pattern echoes upperside. **DISTRIBUTION** Although this is one of our most familiar butterflies, it is a breeding species only in the south, but migrates to the rest of Europe from the Mediterranean region annually, even reaching as far north as the Arctic Circle. **HABITAT AND HABITS** As a strongly migratory species, the Red Admiral has no distinctive habitat, and can be seen almost anywhere in town or countryside. However, especially in late summer, it commonly visits parks and gardens, where it takes nectar from a range of shrubs and herbs in preparation for its return journey to southern Europe. During the spring and summer months immigrant Red Admirals continue to breed in northern Europe, and can be seen in any month of the year. However, populations are at their peak in late summer. The eggs are laid on stinging nettles, and occasionally on other plants in the same family.

Painted Lady ■ *Vanessa cardui*

DESCRIPTION Upperside ground colour buff-orange, sometimes with pinkish tint. This shades to black towards apex of forewing, with an array of white spots enclosed. Irregular black markings on rest of forewing and more obscure ones on hindwing, with row of black spots towards wing margin. Sometimes these have blue centres, and there are small blue spots at rear of hindwings. Underside hindwing has intricate pattern of fawn and creamy-white, with row of large spots towards margin, and is criss-crossed with white veins. Forewing pattern resembles that of upperside, but with a more definite pink tint to ground colour. **DISTRIBUTION** Like the Red Admiral (see p. 103), the Painted Lady can be seen throughout Europe, though it is a full resident only in North Africa and the Middle East, where it spends the winter months. **HABITAT AND HABITS** On arrival from their spring migration northwards, the butterflies disperse to a range of habitats: dunes, heaths, downland, open woodland and urban open spaces. They continue to produce new generations, and are reinforced by continuing immigration. Like the Red Admiral they are likely to be seen in urban parks and gardens in the summer, feeding up before their return migration. However, their numbers fluctuate greatly from one year to the next, and they are usually much less common in northern Europe than is the Red Admiral. The eggs are laid on several thistle species, but also on mallow and, in Sweden, mugwort *Artemisia vulgaris*.

Female, upperside

Underside

Camberwell Beauty ▪ *Nymphalis antiopa*

DESCRIPTION A fine, large butterfly. Upperside rich chocolate-brown with row of bright blue spots close to marginal yellow band. Two pale markings set in narrow black border along leading edge of each forewing. Underside mainly black, with whitish borders.
DISTRIBUTION Distributed throughout Europe including Scandinavia, though scarce in northern and western Norway and Sweden. Absent from Britain and Ireland except as a very rare migrant. **HABITAT AND HABITS** Usually inhabits open woodland, where its soaring flight takes it out of the reach of cameras. Like some other relatives, it occasionally comes down to the forest floor to sip moisture. Adult butterflies can be seen in July–September in the northern part of their range, and again in spring after hibernation. Caterpillars feed on the leaves of sallow (notably *Salix caprea*), as well as other willows, poplars and birch.

Underside

Upperside

Large Tortoiseshell ■ *Nymphalis polychloros*

DESCRIPTION Upperside bright orange-red, with black spots on forewings, and black borders that, on hindwings, include small, triangular blue markings. Underside is in strong contrast, a mixture of black and dark-brown lines and spots, giving excellent camouflage when the wings are closed during hibernation. **DISTRIBUTION** Uncommon but widespread species across Europe, reaching as far north as the south of Norway, Sweden and Finland. In Britain it was once a familiar species in the south-east and East Anglia, but has probably been extinct as a resident there for 50–60 years. There are occasional sightings, probably of immigrant individuals. **HABITAT AND HABITS** Like some other related species, the Large Tortoiseshell has no defined habitat. Butterflies roam the countryside, especially well-wooded districts. Females seek out elms (especially wych elms), or sallows and willows on which to lay their eggs. These are clustered around a twig and the caterpillars feed together. Adults appear in July–August, feed up on garden flowers, sap runs and honeydew before hibernating in trees, wood piles or other sheltered places. They fly again in March–April.

Underside

False Comma ▪ *Nymphalis vau-album*

DESCRIPTION Upperside pattern similar to that of the Large Tortoiseshell (see p. 106), but ground colour is dull yellow-orange to red-brown, dark borders are wider, and there is a white patch close to apex of forewing (as in the Scarce Tortoiseshell, below). Small white spot in middle of underside hindwing. Jagged wing outline similar to that of the Comma (see p. 108), but it is larger and darker in colour. **DISTRIBUTION** Eastern species that is a very rare migrant to parts of north-eastern Europe. **HABITAT AND HABITS** Like the other large tortoiseshells, a butterfly that wanders through the countryside, but is most often seen in lowland woods and wood edges. There is one generation each year, emerging in July–August. After hibernation they fly again in spring, when females lay eggs on twigs of trees or shrubs, such as aspen, elm, ash, birch or sallow.

Scarce Tortoiseshell ▪ *Nymphalis xanthomelas*

DESCRIPTION Upperside very similar to that of the Large Tortoiseshell (see p. 106), but there is white infilling in space between outermost black marking on leading edge of forewing and wing margin. Underside, too, is very similar, but legs are light brown rather than black as in the Large Tortoiseshell (hence the slightly misleading alternative English name 'yellow-legged tortoiseshell'). **DISTRIBUTION** Eastern species with scattered distribution in central and eastern Europe. Immigrant species in most of Europe, but has recently extended its range into the Baltics and southern Scandinavia. Occasional migrants reported in Britain, with a significant influx in 2014. **HABITAT AND HABITS** Inhabits damp lowland deciduous woodland. As in the Large Tortoiseshell, adults frequently bask with their wings open on warm surfaces. Butterflies emerge in July and can be seen until they enter hibernation in September. The eggs are laid in clutches on twigs of willows (*Salix* species) and poplars (*Populus* species), and the resulting caterpillars feed gregariously in their early stages.

Comma ■ *Polygonia c-album*

DESCRIPTION Two distinct colour forms. Dark, 'usual' form has rich orange-red ground colour on upperside, with black spots and dark borders. Underside dark brown with a few greenish lights, and white 'C' on hindwing. The other form, *hutchinsoni*, has a paler, golden upperside with more obscure dark markings. Underside of this form much paler, a complex pattern of browns, fawn and grey, with green lights. In both forms the jagged wing shape is quite distinctive. **DISTRIBUTION** Common and widespread in Europe as far north as southern Scandinavia. In Britain it came close to extinction in the early part of the 19th century, with a recovery in the early to mid-20th century. Now widespread throughout

England and Wales, and expanding northwards in Scotland. Very localized in Ireland. **HABITAT AND HABITS** In spring the butterfly is most likely to be seen on the edges of woods, along hedgerows or in sunny, sheltered field corners, but in summer and early autumn it visits parks and gardens, where it feeds up on flowers or fruit juices before hibernation. In Scandinavia there is generally just one brood, but further south some of the eggs that are laid in spring develop quickly and emerge as *hutchinsoni* adults, while others develop more slowly and emerge as the dark form. Only individuals of the dark form survive to hibernate. Caterpillars feed mostly on hops, stinging nettles or elm.

Form hutchinsoni

Autumn

Form hutchinsoni, *underside*

Small Tortoiseshell ▪ *Aglais urticae*

DESCRIPTION Upperside bright orange-red, with alternating blocks of yellow and black along leading edge of forewing, the outermost patch having yellow replaced by white. Three other black spots in each forewing, and basal area of hindwing is black. Black outer borders of fore- and hindwings enclose series of blue spots. Underside hindwing black with wavy paler band in outer part of wing, while forewing has similar pattern to that of upperside, but with greatly subdued colours. **DISTRIBUTION** One of the most widespread and familiar butterflies of Europe, occurring even in the far north of Scandinavia. Present throughout Britain and Ireland, but suffers great fluctuations in abundance from year to year. **HABITAT AND HABITS** May be encountered almost anywhere, along hedgerows, woodland edges and roadside verges, but especially in parks and gardens. Adult butterflies fly from July into the autumn, when they find holes and crevices in trees or human habitation in which to hibernate. They can be seen again from early spring until May or June. In southern Britain the early summer brood produces a second generation in autumn, but in Scotland and Scandinavia there is usually just one generation in a year. The eggs are laid on stinging nettles.

Underside

Peacock ■ *Aglais io*

DESCRIPTION Upperside strikingly beautiful, with colourful 'eye'-markings against a red-brown ground colour on both fore- and hindwings. Underside has an intricate pattern of black and dark grey patches and lines, with metallic iridescence. **DISTRIBUTION**

Underside

Common and widespread butterfly in Europe, but more localized in the south. In northern Europe it is now common as a resident species in southern and central Scandinavia, where it was, before the 1970s, mainly represented by migrants. Common and widespread throughout Britain and Ireland. **HABITAT AND HABITS** A wanderer with no defined habitat, and can be seen in woodland glades and rides, roadside verges and along hedgerows. However, like the Small Tortoiseshell (see p. 109), it is a frequent visitor to parks and gardens, especially in summer and autumn as it builds up its fat stores before hibernation. When threatened by a predator it flashes its wings and makes a hissing sound by scraping them together. It flies from July into autumn, and again in spring after hibernation. The eggs are laid on stinging nettles and the young caterpillars feed together in a web.

High Brown Fritillary ■ *Argynnis adippe*

DESCRIPTION Upperside ground colour orange with outer row of black spots and zigzag black lines. Underside hindwing has pattern of silver spots, and between the outer two rows of these is a further row of red-ringed silver points. This is absent in the very similar Dark Green Fritillary (see p. 112), which sometimes flies with the High Brown. There is a slight difference in the shape of the forewings, too. In some specimens the silver spots are replaced by pale cream-yellow ones. **DISTRIBUTION** Found throughout most of Europe, but its distribution is more 'patchy' than that of the Silver-washed Fritillary (see p. 115). Occurs northwards to east-central Sweden and still occurs in a few British localities, having declined rapidly since the 1950s. Absent from Ireland. **HABITAT AND HABITS** Habitat includes mature oak woodland with glades, wide rides and, usually, active coppice management. Also occurs in rough, grassy hillsides, often with gorse scrub and bracken. Remains relatively common in southern Scandinavia, but in Britain it declined drastically from the 1950s onwards, and is regarded as the most endangered species there. Now occurs in a very few places in south-western England, south Wales and around Morecambe Bay in the north-west. Concentrated conservation efforts seem to have at least halted its decline. Flight period is from beginning of July to late August, and the eggs are laid on bracken fronds, stones or leaf litter close to vigorous growths of violet.

Underside

Female

Dark Green Fritillary ■ *Argynnis aglaja*

DESCRIPTION Male upperside bright orange with pattern of black spots and irregular black line running across both fore- and hindwings. Female often very similar, but in some strikingly beautiful specimens the orange is paler, and parts of the wings are suffused with dark green. Underside hindwing pale orange with green on basal area, and array of silver spots. Area between two outer rows of silver spots unmarked (compare High Brown Fritillary, see p. 111). **DISTRIBUTION** Occurs almost throughout Europe, except

Mating pair

Mediterranean islands and northern Scandinavia. In Britain occurs up to the north of Scotland, and is also widespread in Wales and Ireland. However, its habitat is quite specialized so it is often very local within that range. **HABITAT AND HABITS** Sometimes seen in open woodland, but its favoured habitats are open rough grassland, heath, moorland and coastal dunes. In suitable habitat can occur in high numbers, and the sight of dozens taking nectar from a patch of marsh thistle is not to be forgotten. However, the unfertilized, flower-rich grasslands it favours have been lost over most of the farmed countryside. Flies from end of June to early August, and the eggs are laid on various species of violet.

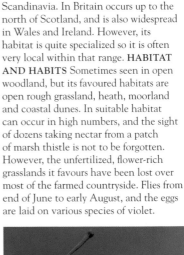

Males

Undersides

Pallas's Fritillary ■ *Argynnis laodice*

DESCRIPTION Upperside slightly more yellowish-orange than in other large fritillaries, and markings around middle of forewing form separate spots, rather than irregular lines. Males have streaks of dark scent scales on two veins only, towards rear of forewing, and overall wing shape is rounded. Underside beautiful and distinctive – yellow-orange in basal half, and flushed with violet scales on outer half, a line of white markings separating the two. **DISTRIBUTION** European distribution is primarily eastern, widespread in Romania and Slovakia, and further north as far as the Baltic states and southern Finland, with isolated records from southern Sweden. Appears to be extending its range northwards. **HABITAT AND HABITS** Inhabits wide rides and glades in humid forests, but also damp, scrubby grassland and woodland edges. Visits wayside flowers such as bugle, and is relatively easy to approach. Flight period is from mid-July to late August, and the main food plant of the caterpillars is marsh violet *Viola palustris*.

Male

Underside

Niobe Fritillary ■ *Argynnis niobe*

DESCRIPTION Upperside very similar to that of both the High Brown and Dark Green Fritillaries (see pp. 111 and 112), though often with darker wing margins. Ground colour of underside hindwing usually pale orange-brown, sometimes suffused with green. Pattern of silver spots and intermediate row of red-ringed silver points very similar to that of the

High Brown, but there is usually a small yellow spot in the cell closest to the body, and the veins are outlined in black (orange-brown in the High Brown). As in the High Brown, the silver spots are sometimes replaced by pale cream-yellow ones, and this is the most common form in Scandinavia. **DISTRIBUTION** Widespread in mainland Europe, though often more localized than the three *Argynnis* species so far described. However, occurs in Belgium, the Netherlands, Denmark, the Baltic states and southern Scandinavia. Absent from Britain and Ireland. **HABITAT AND HABITS** Occurs in flower-rich grassland on sunny slopes, often at high altitudes, and the caterpillars feed on various species of violet. Flight period in the north is in June–August.

Male

Underside

Silver-washed Fritillary ■ *Argynnis paphia*

DESCRIPTION The largest of the fritillaries that occur in northern Europe. Male tawny-orange on upperside, with complex pattern of black spots and irregular lines. On forewing several wing veins are enlarged by broad black lines of scent scales, a distinctive feature of this species. Female has a more subdued ground colour, usually suffused greenish towards the base. A minority form of female (f. *valezina*) has upperside wholly suffused with green. Underside in both sexes has an irregular green pattern on hindwing, crossed by 'washes' of silver. **DISTRIBUTION** Common and widespread throughout most of Europe, reaching as far north as southern Sweden, Norway and Finland. In Britain widespread in the south and west of England, and Wales, but more localized in the east. Widespread in Ireland, but currently absent from northern Britain. In eastern England it is regaining lost ground. **HABITAT AND HABITS** Predominantly a butterfly of woodland glades, wide rides and recently felled compartments in forests, both broadleaved woodland and conifer plantations. In some areas inhabits scrubby meadows. Males are most readily observed as they float effortlessly along a forest ride on patrol for newly emerged females. Both males and females gather to take nectar from bramble patches, where they can usually be easily approached. Female lays her eggs in tiny cracks in tree bark, having first checked for violets nearby. The following spring the caterpillars have to journey down the trunk to find the violets on which they feed. Butterflies are on the wing from beginning of July to early September.

Male

Female

Underside

Queen-of-Spain Fritillary ■ *Issoria lathonia*

DESCRIPTION Males bright orange on upperside, with bold black markings and often greenish tint close to wing bases. Females usually have paler ground colour, and more extensive greenish suffusion over wings. Underside spectacular, with irregular array of large silver spots. In both sexes wing shape is distinctive, outer margin of forewings being markedly concave. **DISTRIBUTION** Highly mobile butterfly that can be encountered almost anywhere in Europe up to southern Scandinavia in the north. Though it is not (yet) resident in Britain, migrants are frequently reported in the south. **HABITAT AND HABITS** Inhabits hot, dry meadows, fallow fields, coastal dunes and rough, scrubby grassland. Easily approached as it takes nectar from low-growing flowers. In northern Europe usually has two generations each year, in May–June and July–August, and the caterpillars are reported to use mainly wild and field pansies *Viola tricolor* and *V. arvensis*, though other species of violet are used elsewhere. Further south in Europe the species may have as many as four generations in a year.

Male

Female

Underside

Marbled Fritillary ■ *Brethnis daphne*

DESCRIPTION Very similar to its smaller relative, the Lesser Marbled Fritillary (see p. 119). Wing shape rounded, the upperside ground colour orange in both sexes, with irregular black lines and two outer rows of black spots on each wing. Underside, too, similar to that of the Lesser Marbled Fritillary. Hindwing has outer orange-brown area with row of black eye-markings and violet flush. Across middle of wing is a band of yellow cells, the yellow of one of the outer cells being obscured by orange suffusion (clear yellow in the Lesser Marbled Fritillary).

Underside

DISTRIBUTION Predominantly a butterfly of south-central and eastern Europe, but has outlying populations in northern Poland and southern Lithuania. Absent from Britain, Ireland and Scandinavia.

HABITAT AND HABITS Habitat is quite varied, including woodland edges, warm hillsides and abandoned cultivation, but always with patches of scrub. Adult butterflies take nectar from a wide range of flowers, but are especially associated with brambles. Flight period is from end of May to beginning of August. Caterpillars feed on the leaves of bramble and some related plants.

Male

Twin-spot Fritillary ■ *Brethnis hecate*

DESCRIPTION Upperside orange with characteristic fritillary pattern of black lines and spots. However, double rows of black spots towards outer margin of each wing run

parallel to one another, and this pattern is repeated on underside. On underside hindwing there is a band of pale cells across the middle, as in the other *Brethnis* species, but linking this with the outer margin there is a pale yellowish wedge that interrupts the mainly orange-red outer area of the wing. **DISTRIBUTION** Butterfly of the south and south-east in Europe, with an outlying northerly population in Lithuania. **HABITAT AND HABITS** Habitat is dry, grassy areas at low to medium altitudes in mountains, often among scrub or light woodland. There is one generation each year, and the flight period is from end of May to end of July. Caterpillars feed on meadowsweet *Filipendula ulmaria*.

Underside

Male

Lesser Marbled Fritillary ■ *Brethnis ino*

DESCRIPTION Wing shape distinctive, being more rounded than in other similar fritillaries. Upperside colour pattern similar to that of *Boloria* group, but black markings are more sparse and there is a distinct border, formed of two narrow black lines that are often (especially in females) fused to form a marginal black band. Underside hindwing distinctive, with series of brown eye-spots and variable dusting of violet scales across middle, and pale fawn band closer to wing base. **DISTRIBUTION** Widespread and not uncommon in most of Europe, reaching as far north as southern Norway and Finland, and southern and eastern Sweden. Absent from Britain and Ireland. **HABITAT AND HABITS** Inhabits damp, flowery grassland, often in woodland clearings, such as cleared tracks under power lines. Flies from mid-June to early August (according to locality), and the eggs are laid on meadow sweet, dropwort (*Filipendula* species) and cloudberry in its northern range.

Underside

Male

Female, Gotland

Marsh Fritillary ■ *Euphydryas aurinia*

DESCRIPTION Very variable butterfly, with several named forms in different parts of
Europe. Usual form in northern Europe has network of black lines with orange-red bands
on both fore- and hindwings, and patches of a similar colour closer to wing bases, with
paler yellow-orange areas across middle of wings. The contrast between these areas on the
wing is usually greater in males than in females. Underside a more subdued repetition of
upperside pattern. Outer orange band on hindwings on both surfaces has black spot in each
cell (Scarce and Lapland Fritillaries, see pp. 122 and 121, lack this). **DISTRIBUTION**
Very widespread but increasingly localized across Europe as far north as southern Finland
and southern Sweden. In Britain limited to western parts of England and Wales, but more

widespread in Scotland and also in Ireland.
HABITAT AND HABITS In northern
Europe inhabits two quite different sorts of
environment. Generally found in lightly
grazed wet pastures, heaths and moors, or
boggy ground close to lakes. However, it
can also thrive on dry, flower-rich chalk and
limestone downland. Males fly low over
their habitat in search of females, which tend
to skulk in vegetation. The heavy-bodied
females crawl over the caterpillars' food
plant, devil's-bit scabious, and on a suitable
plant lay large batches of some 150 eggs. The
resulting caterpillars spin a web and feed
gregariously. Butterflies are on the wing from
middle of May to late June, later in the north.

Underside

Male

Courtship

Lapland Fritillary ■ *Euphydryas iduna*

DESCRIPTION The wing pattern, with its combination of red bands, white central area and black outlining, distinguishes this species from all others in the region.

DISTRIBUTION In Europe occurs only in a restricted area of northern Norway, Sweden and extreme north-west Finland, but distribution also extends further east into Asia.

HABITAT AND HABITS At low altitudes found in open, boggy areas among dwarf birch *Betula nana*, heathers and other low vegetation, but higher up inhabits bare rocky terrain. Within its narrow geographical range it can be quite abundant. However, its numbers fluctuate widely from year to year and it has a very short, weather-dependent flight period (usually from late June to early or mid-July). In dull or wet weather it hides deep in vegetation or among hollows and cracks in rocks, appearing when the sun comes out to bask with open wings until it is warm enough for flight. This is fast and low over vegetation, with males flying up to intercept passing females. The eggs are laid on various plants, including species of *Veronica*, *Vaccinium* and *Bartsia*, and the caterpillars are said to feed communally in webs.

Underside

Female

Scarce Fritillary ■ *Euphydryas maturna*

DESCRIPTION Colour pattern on upperside resembles that of other members of the marsh fritillary group (*Euphydryas*), but is much more strongly contrasting, with extensive black in-filling towards wing bases, and wide orange-red band running along outer borders of both fore and hindwings. Underside more evenly coloured, with bands of orange and cream, and lacking spots. **DISTRIBUTION** As its name implies, this is a scarce and declining butterfly, with scattered populations across Europe including the Baltic states, southern Finland and, very locally, southern Sweden. Absent elsewhere in northern Europe. **HABITAT AND HABITS** Inhabits woodland edges, rides and glades, and can be quite numerous where it occurs. On the wing for about a month from early June. The eggs are laid in dense, layered clusters on the undersides of leaves of ash, guelder rose or cow-wheat, and the caterpillars feed communally.

Male

Female

Underside

Cranberry Fritillary ■ *Boloria aquilonaris*

DESCRIPTION Upperside orange with black markings in male, paler, yellow-orange in female. Towards rear of each forewing are two adjacent black 'V' markings. Underside similar to that of the Mountain Fritillary (see p. 131), but hindwing ground colour is deeper red and pattern more strongly contrasting. Also, pattern of black spots on underside forewing is much more strongly marked (reduced in the Mountain Fritillary). As in the Mountain Fritillary, hindwing is angled rather than smoothly curved. **DISTRIBUTION** In Europe this is primarily a species of the north and east, though extending as far west as France in the southern part of its range. Widespread in northern Europe, reaching as far as northern Norway, but absent from Britain and Ireland. **HABITAT AND HABITS** Lives in wet meadows, marshes and bogs, especially in the vicinity of lakes. Adults emerge in the middle of June and are on the wing until late July. Caterpillars feed on cranberry and bog rosemary *Andromeda polifolia*. Many of its habitats have been destroyed by drainage, especially in the southern part of its range.

Underside

Male

Arctic Fritillary ■ *Boloria chariclea*

DESCRIPTION The butterfly could be confused with the Polar Fritillary (see p. 132) as their distribution and habitats overlap, and they are similar in size and general appearance.

However, black markings on upperside of the Arctic Fritillary are finer, lacking broad central zigzag black band of the Polar Fritillary. Orange ground colour is also brighter, and there is less dark suffusion. The most distinctive feature is the shape of silver-white spots around the margin of the underside hindwing. These are oblong/oval in the Arctic Fritillary, but 'T'-shaped (that is, with a distinct 'stalk') in the Polar Fritillary. **DISTRIBUTION** As its name implies, this is a High Arctic species, in its European range only found in the far north of Scandinavia, with a distribution similar to that of the Polar Fritillary, but extending a little way further south in the mountains of Sweden. **HABITAT AND HABITS** Although the two fritillaries overlap in their preferred habitat, the Arctic Fritillary is more likely to be found lower on the same mountain slopes, in the transitional zone between birch woodland and the open mountain heath. Male flies up to intercept passing fritillaries, and occasionally both sexes can be seen taking nectar from patches of moss campion *Silene acaulis*. Flight period is variable, depending on weather conditions, but starts sometime in mid-June–mid-July. Caterpillars are thought to feed on white arctic bell-heather *Cassiope tetragona*.

Underside

Weaver's Fritillary ■ *Boloria dia*

DESCRIPTION Upperside orange, boldly marked with black spots and lines. Complete row of black spots towards margin of each fore- and hindwing. Underside pale fawn-yellow with reddish-brown patches, and violet flush along with row of dark brown spots in outer part of hindwing. Leading edge of hindwing angled, rather than smoothly curved. **DISTRIBUTION** Found in most of mainland Europe, though absent from much of the Mediterranean region and, in the north, reaches as far as Belgium and Estonia. Absent from Scandinavia, Britain and Ireland. **HABITAT AND HABITS** Inhabits range of grassy locations, including dry meadows, roadsides, hillsides, woodland clearings and scrub. There are two generations each year in the north, but sometimes three in the south, and the butterfly can be seen from May–September. Caterpillars feed on several species of violet.

Female

Male

Underside

Bog Fritillary ■ *Boloria eunomia*

DESCRIPTION Upperside, with its orange ground colour, black lines and outer rows of black spots, is similar to that of other *Boloria* species. However, underside hindwing is distinctive, with row of eye-markings near outer border. **DISTRIBUTION** Occurs in the Alps, Pyrenees and other European mountain ranges, as well as less commonly in northern Germany, Poland and the Baltic republics. Widespread and often common in Norway, Sweden and Finland, but absent from Britain and Ireland. **HABITAT AND HABITS** In the northern part of its range, often abundant in boggy areas among dwarf birch, marsh andromeda and *Vaccinium* species, and associated flowery slopes. Butterflies fly for about a month from mid-June in southern Scandinavia, two or three weeks later in the far north. The eggs are laid on *Vaccinium* and probably other low-growing plants.

Female

Underside

Male

Pearl-bordered Fritillary ■ *Boloria euphrosyne*

DESCRIPTION Adult butterfly has several local variations, but generally orange-brown on upside, with 'standard' *Boloria* pattern of black markings. Underside red-brown with paler markings and row of silver spots around hindwing margin. There is a median yellow band, which has a silver oblong at middle. Cell closest to body has small, yellow-ringed black spot. The Small Pearl-bordered Fritillary (see p. 133) is very similar, but usually has more silver-white spots on underside hindwing and spot is usually larger. In central and northern Scandinavia there is a distinctive form (ssp. *lapponica*) in which upperside markings are more extensive and, especially in females, large areas of wings are suffused with dark green-black scaling. **DISTRIBUTION** Widespread in most of Europe, but absent from southern Iberia. Widespread and common in most of Scandinavia, including the far north. Has declined drastically in Britain, becoming virtually extinct in eastern and central England. Remains fairly widespread in Wales, north-

western England and central Scotland. Also occurs on the Burren in the west of Ireland. **HABITAT AND HABITS** Across most of its range the main habitat is clearings in woodland, especially where coppicing is maintained, and the butterfly can occupy areas within the first year or two after a cut. However, in many areas it also inhabits dry grassland or heaths with gorse or bracken. In Scandinavia ssp. *lapponica* inhabits lowland bogs. Flight period in southern Britain is early May–June, sometimes with a second brood in August. Further north there is just one generation that flies later, in June–July. Caterpillars feed on violets, as well as bog billberry *Vaccinium uliginosum* in Scandinavia.

Underside

Male

Form lapponica

Freija's Fritillary ■ *Boloria freija*

DESCRIPTION Shares with other species in *Boloria* group of fritillaries an upperside wing pattern of rows of black spots and lines on orange ground colour. Underside has complex pattern of red and whitish areas with distinctive black zigzag mark across middle of hindwing. **DISTRIBUTION** Widespread in Sweden, Norway and Finland, becoming more common northwards. Absent further south in Europe, but has an extended distribution in Asia. **HABITAT AND HABITS** Favours peat bogs, often on mountainsides, and has a low, rapid, zigzag flight, settling frequently. In dull weather rests with wings closed. On the wing from early May in the south, but flies well into July in the northern part of its range. Female lays her eggs singly on various pieces of vegetation, including dead grass stems, and the caterpillars feed on a variety of bog plants such as cloudberry, bog-rosemary and cranberry.

Underside

Male

Female

Frigga's Fritillary ■ *Boloria frigga*

DESCRIPTION One of the most beautiful of the fritillaries, especially on underside. Inner half of underside hindwing red with yellow-orange band and white flashes, while outer area is paler and flushed with violet scales. Upperside pattern typical for the *Boloria* group, but this species is more heavily marked than most others in the group, and wing bases are suffused with dark scales, especially on hindwings. Females generally darker than males. **DISTRIBUTION** A truly northern species, with scattered distribution through Scandinavia to the Arctic, and some outposts in the Baltic states. **HABITAT AND HABITS** Typical habitat is open peat bog fringed by light pine and birch woodland, with numerous flowers such as marsh andromeda, bog bean and cloudberry. Flight period is from middle of June or later according to locality, and is quite short. Butterflies take to the wing only during sunny spells, and males fly fast and low in a zigzag flight around the margin of a bog, and far out over wetter areas with sphagnum mosses. They take nectar from flowers such as marsh andromeda, and the eggs are laid on cloudberry *Rubus chamaemorus*.

Male

Underside

Dusky-winged Fritillary ■ *Boloria improba*

DESCRIPTION Upperside wing pattern similar to that of other *Boloria* species, except that ground colour is suffused with dark grey-brown, making its appearance quite distinctive.
DISTRIBUTION European distribution of this rare and elusive species limited to a small area in the far north of Scandinavia.

Underside

HABITAT AND HABITS Habitat is mountain heaths, usually above 500m. Female flies rather reluctantly, spending much of her time crawling among low plants of dwarf birch, mountain heath and *Vaccinium* species, seeming to take nectar from small flowers as she does so. Males make short, low flights, but their dark colouration makes them hard to follow. They take nectar from moss campion *Silene acaulis* and other mountain flowers. They are usually on the wing from early or mid-July into August. Caterpillars are said to feed on several species of willow (*Salix*).

Mountain Fritillary ■ *Boloria napaea*

DESCRIPTION Male has the usual pattern of black markings on orange ground colour, but black markings across middle of fore- and hindwings are more linear than in other *Boloria* species. Female has similar pattern, but wings are suffused with dark greenish scales. Underside hindwing has areas of red and pale yellow-orange, with silver-white spots around margin and in central area. Leading edge of hindwing is angled, rather than gently rounded as in most other *Boloria* species. **DISTRIBUTION** Sharply discontinuous distribution in Europe, occurring in the Pyrenees and Alps, in the south, then in the mountains of Scandinavia, including the far north. Absent from Britain, Ireland and the rest of northern Europe. **HABITAT AND HABITS** In the south the habitat is flowery alpine and subalpine meadows, but in its northern range it is found in bogs and other wetlands at lower altitudes. Flies from late June to late July, and the caterpillars feed on yellow wood violet *Viola biflora* and alpine bistort *Polygonum viviparum*.

Underside

Polar Fritillary ■ *Boloria polaris*

DESCRIPTION Shares the basic *Boloria* upperside pattern of orange-brown ground colour with black transverse markings, but is more heavily marked, with especially prominent zigzag across middle of forewing, and dark grey suffusion towards wing bases. Underside hindwing has areas and bands of bright red interspersed with yellow and silvery-white patches, including prominent, black-outlined, dumbbell-shaped white mark close to leading edge. Distinguished from the similar Arctic Fritillary (see p. 124), with which it sometimes flies, by 'T'-shaped silver-white markings around border of hindwing underside. **DISTRIBUTION** Species of the High Arctic, in Europe occurring only in the far north of Norway, Finland and Sweden. **HABITAT AND HABITS** Found on rocky mountain plateaux, which it shares with the Arctic Grayling (see p. 166), but also said to occur at lower altitudes close to the sea. Males fly very low in a zigzag flight, contouring rocks and low-growing plants, and occasionally settling to bask or take nectar from flowers such as marsh andromeda *Andromeda polifolia*. Females are much less active, and are thought to lay their eggs on mountain avens *Dryas octopetala*, which abounds on lower slopes. Flight period is usually between late June and late July.

Female

Underside

Small Pearl-bordered Fritillary ■ *Boloria selene*

DESCRIPTION Upperside orange with black spots and lines, as is usual in *Boloria* species. Underside very similar to that of the Pearl-bordered Fritillary (see p. 127), with silvery 'pearl' markings along border of hindwing and also in several patches elsewhere. Black spot in cell closest to body is larger than in the Pearl-bordered Fritillary. In northern Scandinavia underside ground colour is paler, and silver is replaced by white.
DISTRIBUTION Very widespread in suitable habitat throughout Europe, except Mediterranean areas across the south. Though widespread in Britain, it is very localized and declining, especially in the south and east. Its remaining strongholds are in Wales, north-western England and Scotland. **HABITAT AND HABITS** Characteristic habitat

is woodland glades, wide rides and coppiced areas, a few years after a cut. However, also occurs on rough grassland with scrub or bracken and, in the north, on moorland and lightly grazed wood pasture, and in northern Scandinavia in birch zone on mountainsides. Males fly low in search of females, which spend much of their time taking nectar from flowers such as bramble, or scattering their eggs over clumps of violets (usually *Viola canina* or *V. palustris*). Flight period is late May to early July in England, later further north, sometimes with a smaller second brood in August.

Underside

Male

Female, Scotland

Thore's Fritillary ▪ *Boloria thore*

DESCRIPTION Northern subspecies (*borealis*) paler and with less dark suffusion on upperside than alpine form. However, black lines and spots near outer wing margins in subspecies *borealis* are still more extensive than in other members of the *Boloria* group, giving the butterfly a darker appearance on the upperside. Underside more even toned than that of other members of the group, having areas of pale orange, a yellowish band and orange spots on outer part of hindwing. Slight dusting of violet scales across middle of wing. Underside forewing yellow-orange with black lines. **DISTRIBUTION** There are two main subspecies of this butterfly, one being rather local in the Alps, the other occurring very much further north in the mountains of Finland, Sweden and Norway. **HABITAT AND HABITS** In the north inhabits damp clearings with dappled shade in birch woodland, and takes nectar from buttercup, cranesbills and other flowers. Further south in Scandinavia inhabits conifer forests. Usually flies in late June–July, and the caterpillars feed on violets (mainly *Viola biflora*).

Underside

Female, f. borealis

Titania's Fritillary ■ *Boloria titania*

DESCRIPTION Upperside has usual pattern of black lines and spots with orange ground colour, but the arrow-shaped markings around wing borders, and chequered white-and-back fringes, are quite distinctive. Underside hindwing has complex pattern of red-and-brown patches and spots, with jagged black line across wing, and flush of violet scales on outer part. Arrow-shaped darker markings close to wing margins. As in Mountain and Cranberry Fritillaries (see pp. 131 and 123), hindwing is strongly angled. **DISTRIBUTION** Very discontinuous distribution in mainland Europe, occurring on the Alps and Balkan mountains, then in Latvia, Estonia and southern Finland, with scattered outposts in between. Absent from Britain and Ireland. **HABITAT AND HABITS** In the northern part of its range inhabits humid meadows, clearings in damp woodland and edges of marshes. Flight period is from mid-June to late July, later in the north, and the caterpillars feed on various species of violet.

Male

Underside

Heath Fritillary ■ *Melitaea athalia*

DESCRIPTION Form most likely to be encountered in northern Europe has rich, uniform orange ground colour on upperside, with even grid of black lines and variable extents of dark suffusion towards bases of wings. On forewings, third marginal lunule from rear of wing is larger than the others. Underside hindwing has cream-white ground colour with outer band of orange lunules and inner orange patches. No black spots in outer orange lunules, and fine marginal band is infilled with same creamy-white as adjacent lunules. **DISTRIBUTION** Most widespread of *Melitaea* group in Europe, occurring in several named subspecies throughout Europe northwards to southern Scandinavia, with a distinct subspecies in the far north. Absent from Ireland, and in Britain rare and confined to a small number of sites in the south-east and south-west of England. **HABITAT AND HABITS** Occupies several different habitats further south in Europe, but is more specialized in England and Scandinavia, flying in warm, sheltered woodland clearings, or in abandoned hay meadows and heathland that is regularly burned. Butterfly is on the wing from middle of May to as late as July, depending on latitude, and sometimes produces a small second brood as late as September. Caterpillars feed on cow wheat, ribwort plantain and various species of speedwell, according to locality. Remaining colonies in Britain survive only due to conservation management of their habitats.

Male

Underside

Nickerl's Fritillary ■ *Melitaea aurelia*

DESCRIPTION Upperside has characteristic orange ground colour and grid of black lines, shared with other *Melitaea* species. Difficult to distinguish in the field from other close relatives, but lattice pattern on upperside forms very even, regular pattern. As in the following species, narrow marginal band on underside hindwing is slightly darker in colour than adjacent lunules. Slightly smaller than Assman's and Heath Fritillaries (see pp. 138 and 136). **DISTRIBUTION** A rather local butterfly of central and eastern Europe, absent from much of the south and west, and reaching as far north as Belgium in the west and the Baltic states in the east. Rare to threatened in northern part of its range. **HABITAT AND HABITS** Favours warm, dry grassland, heaths and mountain slopes, but also peat mosses. Flies in a single annual generation from June to beginning of August. Caterpillars feed communally in their early stages, usually on plantain.

Underside

Female

Assman's Fritillary ■ *Melitaea britomartis*

DESCRIPTION This member of the heath fritillary group is distinguished by the wider dark lines forming the 'lattice' on its upperside. Remaining orange-brown areas more evenly distributed over fore- and hindwings than in the False Heath Fritillary (see p. 140). Underside wing pattern very similar to that of the heath fritillary, but more contrasting, and narrow band around edge of hindwing is slightly darker in colour than

pale cream-coloured lunules next to it. **DISTRIBUTION** This increasingly rare species occurs in parts of central and eastern Europe, and still has small populations in southern Sweden. **HABITAT AND HABITS** Inhabits open tracks and clear-felled areas in woodland where there is rich ground flora, including the food plants of the caterpillar (speedwells and ribwort plantain). Butterflies are on the wing from mid-June to as late as early August, and fly relatively weakly, close to the ground, occasionally taking nectar from buttercups and other flowers.

Underside

Male

Female

Glanville Fritillary ■ *Melitaea cinxia*

DESCRIPTION Upperside ground colour orange-brown (more yellowish in northerly populations), with lattice of darkened wing veins and cross-cutting brown-black lines as in other *Melitaea* species. Row of black spots in cells close to border of each hindwing, usually more prominent in females. Hindwing underside white with bands of orange-yellow, the outer of which has a black spot in each cell. Black borders to these cells are bowed inwards – a distinctive characteristic. **DISTRIBUTION** Occurs throughout most of Europe, becoming more scarce and localized in southern

Scandinavia. In Britain appears to be at the north-western limit of its range, with stable populations only on the Isle of Wight. **HABITAT AND HABITS** Habitat in northern Europe often coastal, where the species is associated with rocky slippages, but it also occurs in dry grassland on limestone rock. At coastal sites on the Isle of Wight makes use of warm, sheltered landslips where ribwort plantain abounds. The eggs are laid among warm rocks or on the food plant. In most parts of its range the caterpillars feed on various species of plantain (*Plantago*), but in Scandinavia they also use spiked speedwell *Veronica spicata*. Caterpillars feed communally in silken webs.

Underside

Male, Sweden

False Heath Fritillary ■ *Melitaea diamina*

DESCRIPTION Basic pattern on upperside similar to that of the Heath Fritillary (see p. 136) and other members of the *Melitaea* genus. However, black outlining of wing veins and wavy cross-markings are widened to give much darker appearance, with orange ground

colour reduced to small spots. Towards wing bases and over much of hindwing the dark suffusion wholly replaces the orange. Underside hindwing has distinctive orange-brown band just within row of marginal pale lunules. Each cell in this band frequently has a small black spot attached to its outer edge, often with a paler 'halo'. **DISTRIBUTION** Occurs widely across mainland Europe, but absent or very localized in the far south. Resident in south-west Finland and southern Sweden. Uncommon in the Baltic states, now extinct in Denmark, and absent from Britain and Ireland. **HABITAT AND HABITS** Inhabits damp, flowery meadows and wide, open areas in forests in Scandinavia. Flies from early June to mid-July and often basks with open wings. The eggs are laid on valerian.

Underside

Male

Spotted Fritillary ▪ *Melitaea didyma*

DESCRIPTION Male upperside brilliant red-orange, with variable, dispersed black spots and squiggles. Fine black marginal line, and chequered black-and-white fringes. Female upperside paler orange than male's, often with suffusion of grey scales on forewings, and more heavily marked with black. Underside hindwing creamy-white with two orange bands, and small black spots and lines. **DISTRIBUTION** Predominantly a species of southern and central Europe, which reaches its northern limits in southern Belgium in the west, and Estonia in the north-east. Absent from Britain, Ireland and Scandinavia. **HABITAT AND HABITS** Favours hot, dry localities rich in wild flowers, such as fallow fields, hillside gullies, roadside verges and track sides, usually at low altitudes. Males bask with open wings, and are easily approached while feeding from flowers. Females are less active, seeming to be burdened by their heavy loads of eggs. There are up to three generations in a year in the south, but usually one, in June and July, further north. Caterpillars feed on a range of low-growing plants including plantains, woundworts and toadflax.

Female

Underside

Male

Knapweed Fritillary ■ *Melitaea phoebe*

DESCRIPTION Slightly larger than most other *Melitaea*, but with similar upperside pattern. Ground colour variable, often yellow with orange bands, but sometimes evenly orange. Third marginal orange lunule from rear of the forewing very large. Underside hindwing with white and orange bands, inner margin of cells in outer orange band curve out towards the base of the wing (contrast Glanville Fritillary (see p.139)).

Underside

DISTRIBUTION Widespread and often common throughout southern and central Europe reaching as far north as the Baltic states, but absent from Britain, Ireland and Scandinavia. Now distinguished from very similar *Melitaea telona*, which coexists with *phoebe* in southern Europe. **HABITAT AND HABITS** Inhabits a wide range of flowery grasslands, including light scrub and open woodland. Adults take nectar from a wide range of flowers, especially knapweeds and scabious in the summer, often basking with open wings. They are on the wing from June to late August in northern part of range, with two broods each year in the south. The caterpillars feed on knapweeds.

Male

Female

SATYRINAE – BROWNS, RINGLETS, GRAYLINGS & MARBLED WHITES

Members of this subfamily lack the gaudy colour patterns of the Nymphalinae, most of them having predominantly subdued brown or black-and-brown colour patterns, though many have varying amounts of orange, especially on the uppersides, and one group, the marbled whites, has a black-and-white chequered pattern. Almost all species have prominent 'eye' markings near the apex of the forewings. The ringlets in the genus *Erebia* form a large group of mainly mountain butterflies in Europe. Only a few of these occur in the north, some of them closely related to ones that fly in the Alps. Most of the European Satyrinae use grasses of various species as the food plants of their caterpillars.

The Large Wall Brown, Marbled White, Meadow Brown, Woodland Brown, graylings and most of the northern *Erebias* are fairly large with wingspans of 4.5–5cm. Slightly smaller are the Wall Brown, Speckled Wood, Gatekeeper, Ringlet, Northern Wall Brown and woodland ringlets, at 3.8–4cm. The Mountain Ringlet is smaller still, at 3–3.5cm. The heaths (*Coenonympha*) are all small butterflies, with three species averaging 2.8–3.3cm, the Large Heath and Scarce Heath being larger at about 3.7–3.9cm.

Speckled Wood ▪ *Pararge aegeria*

DESCRIPTION In the north, upperside is dark brown with pale cream-yellow spots and white-centred eye-spot near to forewing apex. Two or three yellow-ringed eye-spots around margins of hindwings. Male has broad area of dark scent scales across forewing. Underside hindwing fawn shading to red-brown, with brown zigzag lines and row of small eye-spots close to margin. Underside forewing echoes pattern on upperside. Further south in mainland Europe pale spots are orange, giving the butterflies a quite different appearance. **DISTRIBUTION** Common and widespread throughout southern and central Europe, reaching as far north as southern Scandinavia. In Britain became scarce in the 19th century, but has recovered since the 1930s and is now generally widespread. Widespread in Ireland. **HABITAT AND HABITS** Characteristic habitat is mature deciduous woodland with sheltered glades and rides, but also found in conifer plantations, scrubland, parks and gardens, and around tall hedgerows. Some males lie in wait for females, occupying perches in sun-spots, while others seek them out by patrolling woodland rides and glades. In Britain and southern Scandinavia there are two or three generations in a year, so the butterfly can be seen in March–October. In the north there is just one generation, in late May to late June. The eggs are laid on the leaves of several grass species.

Male showing underside

Large Wall Brown ■ *Lasiommata maera*

DESCRIPTION In the south of Europe both sexes have extensive areas of orange on upperside, but the northern form (f. *borealis*) is rather smaller, and upperside of male is dark brown, with a prominent, orange-ringed eye-spot at apex of forewing, and three or four similar spots close to margin of hindwing. Female usually has more orange on forewing. Underside silvery-grey with wavy darker lines, an uneven row of eye-markings on hindwing and a large eye-marking at apex of forewing. **DISTRIBUTION** Much more widespread across most of Europe than the Northern Wall Brown (see p. 146), but in Scandinavia it is mainly southern. Occurs in the Baltic states, Poland, the Netherlands and Belgium, but absent from Denmark, Ireland and Britain. **HABITAT AND HABITS** Inhabits open, grassy areas with bare ground or rocky outcrops, where it rests with wings wide open in the sun. In southern Europe has two generations each year, but in the northern range there is only one, from mid-June to mid-July. The eggs are laid on a wide variety of grass species.

Female

Male

Underside

Wall Brown ■ *Lasiommata megera*

DESCRIPTION Upperside orange with brown irregular lines and white-pupilled eye-markings towards apex of forewing. Hindwing has row of large orange spots enclosing three or four eye-markings. Underside hindwing grey with jagged brown cross-markings and arc of eye-markings around margin. Underside forewing has large, white-centred eye-marking close to apex. **DISTRIBUTION** Widespread and often common across Europe, including mainly southern coastal areas of Norway and Sweden. Also occurs in the southern Baltics, eastern Denmark, and Britain and Ireland. In Britain has declined dramatically in recent decades inland, but persists in coastal areas and the south-west. At the same time it is extending its range northwards into Scotland. **HABITAT AND HABITS** Inhabits dry,

sparse unfertilized grassland, downs, heaths and dunes with bare patches. Males are most often noticed, as they bask in warm, exposed spots while lying in wait for passing females. Courtship is described as 'rumbustious'. Except when occupied in this way, males are very difficult to approach, as they quickly fly up, then settle a few metres away. There are two generations a year in the northern part of the range, from late April to early June, then in August–September. Further south early-emerging summer butterflies go on to produce a third brood that flies in the autumn. Eggs are laid on various grasses.

Underside

Male

Female

Northern Wall Brown ■ *Lasiommata petropolitana*

DESCRIPTION In both males and females, upperside ground colour is dark brown, with prominent eye-markings at apex of forewing, and row of 3–4 smaller ones near edge of each hindwing. Eye-spots usually set in orange patches that are more extensive in females than in males. Faint, wavy darker lines across forewing, and another line across middle of hindwing. Underside ground colour paler with greyish suffusion, and markings more clearly defined than on upperside.

Female

DISTRIBUTION Occurs in mountain ranges of southern and central Europe, but in northern range is confined to some areas in the Baltic states, Sweden, Finland and southern Norway. **HABITAT AND HABITS** Favoured habitat is wide, bare tracks and other open places in woodland, where the butterflies settle on the ground or on rocks and bask with open wings, which they close when approached. They take nectar from flowers such as meadow cranesbill. Butterflies are on the wing from early May to late June, or later in the north. Caterpillars feed on various species of grass.

Male

Underside

Woodland Brown ■ *Lopinga achine*

DESCRIPTION Upperside dark brown with rows of large, yellow-ringed darker brown spots or eye-markings towards edges of wings, and faint arc of yellow from leading edge of each forewing. Underside is especially striking, with row of prominent yellow-ringed eye-markings within white band on hindwing.

DISTRIBUTION Widespread but uncommon in woodland across central and eastern Europe, reaching north in north-east Poland, the Baltic states, southern Finland and the south-eastern fringe of Sweden, including Gotland. Absent from Britain, Ireland and Denmark. **HABITAT AND HABITS** Inhabits clearings and grassy verges alongside broadleaved or conifer woodland, where it is surprisingly difficult to follow with the eye as it flies in and out of dappled shade. Flight period is from late June to end of July. Caterpillars feed on various species of grass and sedges (*Carex montana* in Sweden). The species seems to be in decline throughout Europe, possibly because of changes in forestry practices.

Male

Underside

Gatekeeper ■ *Pyronia tithonus*

DESCRIPTION Upperside orange with broad brown borders to wings and double-pupiled eye-spot close to forewing apex. Broad black diagonal band of scent scales on male forewing. Underside hindwing diffusely mottled brown, with pale band and incomplete row of small eye-spots. Forewing repeats upperside pattern. **DISTRIBUTION** Common in southern, central and western Europe, but absent from the Baltic states, Denmark and Scandinavia. In Britain common in the south and in the Midlands, and is extending

its range northwards towards the Scottish border. Mainly southern in Ireland **HABITAT AND HABITS** As its English name suggests, this is a butterfly of field edges and hedgerows, but it also occurs in scrubby habitats, woodland glades and edges and, sometimes, gardens. Adults are on the wing from late June to early September, with a peak in early August. There is just one generation each year, and females lay their eggs on the blades of grasses such as common couch, bents (*Agrostis* species) and meadow grasses (*Poa* species).

Male

Female

Underside

Meadow Brown ■ *Maniola jurtina*

DESCRIPTION Female upperside medium brown, with orange patches on forewing, enclosing single, white-pupiled eye-spot close to apex of each forewing. Male much darker than female, with area of black scent scales on forewing. Eye-spot near apex of each forewing, surrounded by much smaller area of orange than in female. Underside hindwing of female pale brown with paler band across middle of wing, which encloses a variable number of very small spots. Prominent eye-mark at apex of forewing. Male underside similar, but hindwing is more uniformly coloured. **DISTRIBUTION** Very widespread and common throughout Europe as far north as southern Sweden, Norway and Finland. Common and widespread in Britain and Ireland, though more localized in the north of Scotland. **HABITAT AND HABITS** As its name suggests this is a grassland species, which must have suffered greatly from agricultural intensification. Despite this it remains very common on downland, fallow fields, hedgerows, roadside verges and woodland rides, and can survive in quite small, neglected patches of grassland. Butterflies have a rather lazy, flapping flight, and are active even in quite dull and cool weather. They often bask with open wings, but close them or fly off when disturbed. They are on the wing in June–August. Caterpillars feed on a range of grasses, including meadow grasses, bents and rye grass.

Male

Male, underside

Female

Dusky Meadow Brown ■ *Hyponephele lycaon*

DESCRIPTION Male upperside golden-brown with narrow, oblique band of scent scales and black apical spot on forewing. Female has well-defined orange band on forewing, enclosing two large black spots. Underside hindwing pale brown with small, irregular black flecks, and irregular dark line across middle, with paler colouration beyond it. There are

no spots. Underside forewing orange with apical eye-marking, and a second eye-marking towards rear of wing in females. **DISTRIBUTION** Mainly a southern species in western Europe, but reaches further north in the east, including the Baltic states and southern Finland (where it has recently recolonized having become extinct in the 1940s). Absent from Britain and Ireland. **HABITAT AND HABITS** Favours dry, hot, sparse grassland, often coastal, and south-facing stony or sandy slopes. Butterfly always settles with its wings closed, so field identification relies on underside features. In its northern range flies from late June or later, through July and into early August. Caterpillars feed on sheep's fescue, red fescue, upright brome and other grass species.

Male, underside

Female, underside

Ringlet ■ *Aphantopus hyperantus*

DESCRIPTION Male upperside very dark brown, and usually with small, inconspicuous eye-markings, though these are sometime absent. Female slightly paler brown, and usually with clearly defined eye-markings on both fore- and hindwings. Underside very distinctive, grey-brown with row of prominent yellow-ringed eye-markings on forewing, and another row of five eye-spots on hindwing, two of them displaced inwards. In flight looks similar to the Meadow Brown (see p. 149), but when fresh has distinctive white fringes to wings. **DISTRIBUTION** Common butterfly throughout most of Europe, except for large parts of the Mediterranean region. In the north reaches as far as southern Norway, southern and eastern Sweden and southern Norway. Occurs throughout Britain and Ireland, but more localized in the north. **HABITAT AND HABITS** Butterfly of damp grassland, often close to woodland edge or scrub. Both sexes bask with open wings in the morning, or after a cool, cloudy spell, but spend much of their time with wings closed. They fly from mid-June to middle of August. Caterpillars feed on a wide variety of common grasses.

Underside

Male

Female

Pearly Heath ■ *Coenonympha arcania*

DESCRIPTION Upperside forewings pale orange-brown with darker borders, while hindwings are dark brown with narrow yellow marginal line at rear. Butterfly almost always settles with wings closed, so identification in the field depends on features of underside.

Hindwing underside is brown, with broad white band in outer half. There are 4–5 yellow-ringed eye-markings along outer edge of this, and another one on inner edge. Underside forewing orange with small eye-mark at apex. **DISTRIBUTION** Common and widespread in much of mainland Europe, reaching as far north as southern Sweden and Norway. Absent from Britain and Ireland. **HABITAT AND HABITS** Inhabits meadows and grassy, open spaces in woodland, forest tracks and roadside verges, and flies in early June–July. The eggs are laid on melic grass *Melica nutans* and probably also on other grasses.

Male

Chestnut Heath ■ *Coenonympha glycerion*

DESCRIPTION Male upperside dark grey-brown, almost completely unmarked, while female has yellow-ginger forewings and dark brown hindwings. Narrow orange marginal band on female hindwing, often with faint eye-markings. Underside hindwing grey, with one or more small white patches, row of six yellow-ringed and blue-white centred eye-markings, and narrow orange marginal band. Underside forewing orange with grey border

and sometimes faint apical spot. **DISTRIBUTION** Widespread and often common in central and eastern Europe, with more scattered populations in the west and south. To the north reaches as far as the Baltic states and southern Finland, where it is currently extending its range. **HABITAT AND HABITS** Inhabits flowery grassland, including mountain slopes, fallow fields and clearings in light woodland. One generation each year, with adults on the wing from early June or later in the northern part of its range, until early August. Like other species in the 'heath' group, they settle with wings closed. Caterpillars feed on various grasses, including tufted hair-grass, bromes, false brome and fescues.

Scarce Heath ▪ *Coenonympha hero*

DESCRIPTION Upperside dark brown with obscure eye-markings. Underside hindwing also dark brown (orange-tinted in females), with white band in outer part of wing, subtending row of 5–6 red-ringed eye-spots. Between these and wing edge are a fine silvery line and continuous orange-red marginal band. **DISTRIBUTION** Very scarce and localized species in Belgium, the Netherlands, northern France and Denmark, and rather more widespread in south-eastern Sweden and across the border into Norway. Absent from Britain and Ireland. **HABITAT AND HABITS** Inhabits open, grassy glades and rides in woodland, as well as damp meadows. Butterflies settle with wings closed and forewing raised, among grasses, or sometimes on adjacent broadleaved plants. They are very easily disturbed, and then 'contour' low shrubs and small trees, or fly up into trees or dense vegetation. Caterpillars probably feed on several grass species.

Part upperside

Female

Male

Small Heath ■ *Coenonympha pamphilus*

DESCRIPTION Upperside ginger-orange in males, more yellow-orange in females, with narrow darkened borders, and faint dark apical spot in forewing. Underside hindwing grey-brown basally, with wavy boundary followed by pale whitish area, then pale grey-brown towards border, usually with small, faint eye-spots. **DISTRIBUTION** Very widespread and usually common throughout Europe northwards to southern Norway, south-eastern Sweden and southern Finland. Widespread and usually common throughout Britain and Ireland, though becoming less common within its range in Britain. **HABITAT AND HABITS** Habitat is almost all types of open grassland with flowers, including lightly grazed pastures, hay meadows, roadside verges, coastal dunes, downland and brownfield sites. In the north there is usually just one generation each year, with adults flying in late May to late June, but further south there are usually two broods, the second one flying in July and August. In some summers there is a third, autumn brood. Butterflies always settle with wings closed. Males gather in leks to attract females, and the eggs are laid on sheep's fescue, meadow grasses (*Poa* species) and other wild grasses.

Britain

Southern Sweden

Large Heath ■ *Coenonympha tullia*

DESCRIPTION Very variable, both individually and geographically. In the north of its range, upperside ground colour is similar to that of the Small Heath (see p. 154), but with darker shades on hindwing. Underside hindwing grey with irregular patch of white, similar to that of the Small Heath. Further south upperside ground colour is darker, especially in male, and eye-markings are more distinct. Underside of southern forms can be particularly well marked, with row of six yellow-ringed eye-markings towards margin of hindwing, and one or more prominent eye-markings on forewing. **DISTRIBUTION** Widespread in eastern, central and northern Europe, but populations are rather localized because of its specialized habitat requirements. Local or absent in much of western and northern Scandinavia. Fairly widespread in northern England, north Wales and Scotland in suitable habitat, as is the case in Ireland. **HABITAT AND HABITS** Wetland butterfly found in peat bogs and mosses, and the margins of mires and damp moorland, with scattered trees, cotton grass, sedges and other wetland plants. Butterflies settle with wings closed, and take nectar from cross-leaved heath and other flowers. Flight period is from mid- or late June, depending on locality, to early August. In Britain the main food plant of the caterpillars is said to be hare's tail cotton grass, but elsewhere caterpillars also feed on beak-sedges, purple moor-grass and various sedges (*Carex* species). Caterpillars may take two years to complete their development.

Upperside

Male, northern England

Female

Scotch Argus ■ *Erebia aethiops*

DESCRIPTION Male upperside black-brown, with broken red band around wing margins enclosing a variable number of eye-spots. Female dark brown with orange-red bands and more prominent eye-spots, especially towards apex of forewings. Male underside dark red-brown, with obscure paler band across hindwing and red band with eye-spots on forewing. Female underside has paler brown ground colour, a strongly contrasting pale band across hindwing and eye-spots in an orange band on forewing. **DISTRIBUTION** In mainland Europe a mainly central and south-eastern distribution, reaching northwards to Latvia and Lithuania, but not extending to Scandinavia. However, it is well established

in northern Britain, occurring in the Lake District and more extensively in Scotland. **HABITAT AND HABITS** Inhabits woodland rides and margins, and sheltered grassland from sea level to about 500m in Britain, but at higher altitudes further south in its range. Butterflies bask with open wings in sunshine, and males either perch to intercept passing females, or patrol tussocks of grass. There is just one generation each year, flying through August. Caterpillars feed on purple moor-grass as well as sheep's fescue, tufted and wavy hair-grasses, and other common grasses. Some colonies may have been lost as a result of climate warming.

Female

Male, underside

Female, underside

Arctic Ringlet ■ *Erebia disa*

DESCRIPTION Upperside dark brown with irregular line of small eye-markings set in orange towards outer margin of forewing. Upperside hindwing plain brown, sometimes with black-centred orange patch at rear. Like many ringlet (*Erebia*) butterflies, both sexes tend to settle with wings closed, showing the distinctive grey-dusted underside hindwing, with a brown central band and thin wavy dark line towards the outer edge. **DISTRIBUTION** Rather localized mountain species of Norway, Sweden and Finland, becoming more common and widespread towards the northern part of its range. **HABITAT AND HABITS** Habitat is bogs and marshes in scrubby heathland or pine forest. Flight period is from late June to mid-July, and males can be seen flying powerfully over boggy ground, disappearing into deep vegetation as soon as clouds obscure the sun. Little is known about the early stages in the life-cycle of this species.

Underside

Lapland Ringlet ■ *Erebia embla*

DESCRIPTION As in other *Erebia* species, upperside ground colour is dark brown, and there is a row of large, yellow-ringed darker spots on both fore- and hindwings, the spot in the forewing apex having two very small white 'pupils'. Underside hindwing dark chocolate-brown with dusting of grey scales towards margins, and row of small, inconspicuous black spots. Underside forewing has row of black eye-markings. **DISTRIBUTION** As its common name implies, an exclusively northern species, widespread but declining in Sweden and Finland, and very locally surviving in southern Norway, Estonia and Latvia. **HABITAT AND HABITS** Inhabits extensive boggy areas within pine forests. Typically bog pools are choked with mosses and sedges, and fringed with dwarf birch, and the species has suffered due to drainage of its habitats. Butterflies fly fast, often high up in the trees, occasionally settling briefly on low vegetation or on a pine trunk with wings closed. The life-cycle takes two years and the butterfly is more common in even than odd years. Flies in mid-June–mid-July, and the caterpillars feed on sedges (*Carex* species).

Mountain Ringlet ▪ *Erebia epiphron*

DESCRIPTION Male upperside dark red-brown, with ill-defined row of marginal reddish patches that enclose varying numbers of small black points. Female similar to male, but ground colour is lighter brown, and she has a more extensive reddish band close to margins of fore- and hindwings, enclosing a complete and regular band of black spots. Underside hindwing mid-brown with marginal row of orange-ringed black spots, while forewing is reddish with marginal orange band enclosing black spots. **DISTRIBUTION** Very scattered distribution in Europe, being found on most of the main mountain ranges. Absent from most of northern Europe, except Britain. There it survives as a distinct subspecies (ssp. *mnemon*) in the mountains of the Lake District and in Scotland, mainly in the Grampian mountains. **HABITAT AND HABITS** Further south inhabits moorland and forest clearings, but in Britain found on open, grassy mountain slopes, often associated

with marshy hollows and rocky outcrops. Butterflies form quite discrete local colonies with little exchange between them, but can be very numerous within a small area. Males need to bask in sunshine before they are warm enough to fly, but then can continue to fly in search of skulking females even when the sun goes in. Females are heavily laden with eggs and rarely take to the wing, though they and males take nectar from tormentil or wild thyme. Caterpillars feed on mat grass *Nardus stricta* in Britain, but use other grasses further south.

Female

Male

Underside

Arran Brown ■ *Erebia ligea*

DESCRIPTION Male upperside very dark brown with red marginal band enclosing row of white-pupiled eye-spots. Female similar to male, but ground colour is lighter, and the bands wider and orange-red. Both sexes have distinctive black-and-white chequered wing fringes. Underside hindwing dark brown (paler in female), with row of orange-ringed marginal eye-spots. Both sexes have pure white streak from leading edge of hindwing. **DISTRIBUTION** European distribution is discontinuous. Occupies a large area of central and south-eastern Europe. Also found in the north and north-east, including the Baltic states and most of Finland, Norway and Sweden, but becomes more localized in the far north. Absent from Britain and Ireland. **HABITAT AND HABITS** Typical habitat is clearings and wide rides in mixed or coniferous forests, and in birch forests in the far north. Butterflies are on the wing from mid-July–mid-August, but have a two-year reproductive cycle that has resulted in their being more common in odd than in even years. Caterpillars feed on a variety of wild grasses, including millet, tufted hair-grass, cock's-foot and purple moor-grass.

Female

Female, underside

Woodland Ringlet ■ *Erebia medusa*

DESCRIPTION Male upperside dark brown with row of eye-markings, set in orange-red patches, towards margin of each wing. Two of these spots near apex of forewing are prominent, and there are usually three well-marked spots on hindwing. Female upperside similarly marked, but ground colour is lighter brown, and orange areas around spots are paler than in male. Underside of male very dark brown, shading to black, with upperside pattern of black eye-spots and orange patches repeated. Female underside similar, but paler. Wing fringes are brown. Very similar to the Arctic Woodland Ringlet (see p. 162), to which it may be closely related. **DISTRIBUTION** Widespread in central and eastern Europe, with the north-western edge of its range just reaching Belgium and Luxemburg, where it is considered vulnerable. **HABITAT AND HABITS** Along with most species of *Erebia*, favours grassy habitats. These may be damp mountain slopes and moorland, but also, as the English name implies, it may be found in light woodland, especially at low altitudes. It is quite easy to approach, and can be seen from May to late July, depending on latitude and altitude. Caterpillars feed on various grass species.

Male, underside

Male

Female

Dewy Ringlet ■ *Erebia pandrose*

DESCRIPTION Ground colour of upperside dark brown, with faint darker line across middle of forewing, and outer reddish area with small dark spots. Silvery-grey underside hindwing of male, with wavy dark line across it, is distinctive. **DISTRIBUTION** Occurs in the Pyrenees, Alps and mountains of southern and central Europe, as well as being widespread in Norway, western and northern Sweden, and Finland. Absent from the British Isles. **HABITAT AND HABITS** Inhabits open, grassy habitats, often among low scrub in mountains, but at much lower altitudes in the north of its range. There, it is found in open heaths with rocky outcrops, bare ground and low-growing herbs. Well adapted to cold conditions, and warms up quickly in the morning sunshine. Basks with open wings, but closes them in dull weather. Flight period is from late June to early August. Caterpillars feed on sheep's fescue grass *Festuca ovina* and probably other grasses.

Male

Female

Underside

Arctic Woodland Ringlet ■ *Erebia polaris*

DESCRIPTION Upperside dark chestnut-brown with double orange-ringed eye-marking near apex of forewing, and often one or two more dark spots in small orange patches below that. Hindwings have row of three or four orange-ringed black spots (sometimes

with minute white 'pupils') close to margin. Underside similarly coloured, the hindwing often having a paler outer band with a row of very small eye-markings. **DISTRIBUTION** In Europe confined to the north of Scandinavia, within the Arctic Circle. **HABITAT AND HABITS** Inhabits flat, dry, sparsely vegetated areas with bare rocky ground, at low altitudes. Can fly in the absence of full sunshine, but soon basks on bare ground with wings open. In cool or cloudy conditions it nestles down in low vegetation or among rocks. Short flight period from mid- or late June into July, and the caterpillars feed on sheep's fescue grass *Festuca ovina*.

Male

Rock Grayling ■ *Hipparchia alcyone*

DESCRIPTION Upperside black-brown with yellowish white band across both fore- and hindwings. This encloses an apical eye-spot and another smaller spot in forewing. Underside hindwing brown with black flecks and irregular black lines, and white band with indefinite outer boundary across middle. Forewing similarly coloured, with prominent

apical eye-spot. **DISTRIBUTION** Fairly widespread in mainland Europe, but localized within its range. Reaches as far north as northern Germany, Poland and Lithuania, and has an isolated outpost in southern Norway. Absent from Britain and Ireland. **HABITAT AND HABITS** Inhabits dry, rocky mountain slopes, as well as clearings and margins in pine forests in Norway. Butterflies settle with wings closed, often on bare ground or rocks, as the English name suggests. The underside pattern gives it effective camouflage, but the raised forewing displays the distracting eye-spot to a potential predator. Flight period is from mid-June–mid-August, and the caterpillars feed on sheep's fescue, Yorkshire fog, Tor grass and probably other grasses.

Marbled White ■ *Melanargia galathea*

DESCRIPTION This strikingly beautiful butterfly is quite unlike the other species in its family (the 'browns'). Brightly chequered black-and-white upperside distinguishes it from all other species that fly in northern Europe. There are small, obscure eye-markings, especially enclosed in the black hindwing borders. Underside markings more subtle, with pattern of black lines, with grey infilling, an irregular band across hindwing, and small eye-markings on both fore- and hindwings. Pale colouring on upperside may be cream or pale yellowish instead of pure white. **DISTRIBUTION** Common over much of southern and central Europe, but absent from the south of the Iberian peninsula. In the north, absent from Scandinavia except as a rare migrant. In Britain occurs in central and southern England, and in the north-east, but absent from Scotland. Also not found in Ireland. **HABITAT AND HABITS** Butterfly of tall grassland with wild flowers, often on south-facing calcareous downland, but also on roadside verges, neglected field corners and abandoned brownfield sites. Butterflies fly from late June to mid-August, reaching a peak around the middle of July. They are easily approached as they bask, wings open, or feed from scabious or knapweed flowers. Caterpillars feed on red fescue, sheep's fescue, timothy, cock's-foot and other grasses.

Underside

Male

Grayling ■ *Hipparchia semele*

DESCRIPTION Male upperside dark brown with two widely spaced eye-spots set in incomplete orange band. Orange marginal band on hindwing, enclosing one or more

small eye-spots, and forewing has darkened band of scent scales. Female upperside similar, but ground colour is a little lighter, there is yellow rather than orange around forewing eye-spots, and there is no band of scent scales. Underside is irrorated dark grey, with black zigzags and paler central band on hindwing, while forewing is yellow or orange with eye-spots. **DISTRIBUTION** Common in most of Europe, though in the south there is some uncertainty about its relationship to very similar forms. In the north reaches as far as southern Norway, Sweden and Finland. In Britain mainly coastal, with remaining inland populations in south-central England and East Anglia. In Ireland mainly coastal. **HABITAT AND HABITS** Typical habitat is south-facing, dry, rabbit-grazed downland or heathland with dry, nutrient-poor soil. In recent decades many populations have been lost from downland, and it now thrives on coastal dunes and lowland heaths. Butterfly always settles with wings closed, initially with the bright underside forewings exposed. It then drops the forewings within the hindwings and is almost perfectly camouflaged. If alarmed, the forewings are raised again, presumably as a defensive display. Butterflies are on the wing from late June or early July to mid-September. Caterpillars feed on grasses such as sheep's fescue, bristle bent and other grasses, notably marram in many coastal sites.

Male

Female

Tree Grayling ■ *Hipparchia statilinus*

DESCRIPTION Male upperside dark grey-brown with two widely separated, faint black spots towards outer margin of forewing, and obscure paler areas on outer parts of both

Male, underside

fore- and hindwings. Ground colour of female lighter brown with similar though clearer patterns than those of male. Both sexes have two white spots between large black ones on forewing. Male underside hindwing finely etched in brown, with irregular dark brown line across middle. There is then a whitish band with brown-and-grey bands towards wing margin. Forewing similarly divided, but with large, orange-ringed black eye-spots and intermediate white spots, as on upperside. Female underside similar, but hindwing is more evenly coloured. **DISTRIBUTION** Mainland European distribution mainly southern and eastern, with some outlying populations in the north-west and north-east. Rare in the northern part of its range, in the Netherlands and Lithuania, and absent from Scandinavia, Britain and Ireland. **HABITAT AND HABITS** Habitat is dry, rocky slopes, sheltered scrubby heathland and sometimes open pine forests. Butterflies invariably settle with their wings closed, usually displaying their prominent eye-markings briefly after landing. Unless alarmed, they sit at rest with the forewings hidden between the hindwings, so that the cryptic patterning of the hindwing makes them almost invisible against the rock, bare ground or tree-trunk on which they settle. They have one generation each year, and are most likely to be seen in July or August. Caterpillars feed on a range of grass species.

Female, underside

Arctic Grayling ■ *Oeneis bore*

DESCRIPTION Upperside pale yellow- or grey-brown, usually without spots, but occasionally with small eye-markings at apex of forewing. Underside ground colour

similar, with darker cross-markings on hindwing forming wave pattern. Wing veins outlined whitish. Often a tiny, inconspicuous white apical spot towards apex of forewing underside. **DISTRIBUTION** A truly Arctic species that occurs in Europe only in the far north of Scandinavia. **HABITAT AND HABITS** Inhabits seemingly inhospitable mountain plateaux, among bare, lichen-clad rocks with *Sedum* and stunted birch *Betula nana*. Rests among rocks with wings closed, occasionally making short, low flights, and on the wing for about two weeks between mid-June and mid-July. Caterpillars feed on sheep's fescue grass *Festuca ovina*, and the life-cycle takes two years.

Male, underside

Norse Grayling ■ *Oeneis norna*

DESCRIPTION Upperside ground colour usually pale yellowish-brown, with variable number of small eye-markings on both fore- and hindwings. Underside forewing usually has one or more of these set against the conspicuous pale orange ground colour.

DISTRIBUTION European distribution limited to Norway, Sweden and the north of Finland, favouring higher altitudes in the southern part of range in Norway. **HABITAT AND HABITS** Flies in clearings in birch forest, on boggy ground among dwarf birch scrub, and in bare, rocky terrain in the north. In common with other graylings it usually settles with its wings closed, but with the forewing underside visible. When fully at rest or when startled, the forewings are lowered so that only the cryptic hindwing undersides are visible. Butterflies are on the wing in mid-June–mid-July, and caterpillars feed on sedges and probably also on coarse grasses.

Baltic Grayling ■ *Oeneis jutta*

DESCRIPTION Upperside grey-brown, darker than in Norse and Arctic Graylings (see p. 166), with variable dark oval spots set in yellow bands or patches close to wing margins. Underside hindwing silver-grey with irregular darker band running across it, and usually one eye-spot set in paler outer band. Underside forewing brown with dark spots in outer yellow band, and grey apex and outer border. The chequered fringes are quite noticeable. **DISTRIBUTION** Found in the Baltic states, southern Sweden and Norway, becoming more localized further north in Scandinavia. **HABITAT AND HABITS** Typical habitat is wide, open peat bogs dominated by bog cotton in pine forests. Males fly from one pine trunk to another, fluttering up from the bottom, then flying down to the base of the next tree, in search of resting females. Courtship and mating take place on the tree-trunks, but the pair then fly off into bog vegetation. Females fly out into the open bog to lay their eggs on bog cotton (*Eriophorum* species). The life-cycle takes two years, and the butterfly appears only in even years. Flight period is from early June to early July.

Part upperside

Female

Male

■ Taxonomic List of Species ■

The classification of butterflies at levels above species is in considerable flux. In some cases there is disagreement among authors as to the allocation of some species to genera, and at the level of subfamilies there is also no consistency in recent literature. In what follows I conform to the allocations of species to genera in Haahtela *et al.* 2011, and dispense with the subfamily level except in the case of the major division of the Nymphalidae into Nymphalinae and Satyrinae (which were formerly allocated to distinct families). Where there are several species in a genus, they are listed in alphabetical order of their specific scientific names.

Family Hesperiidae

Genus *Erynnis*
Erynnis tages Dingy Skipper

Genus *Carcharodus*
Carcharodus alceae Mallow Skipper
Carcharodus flocciferus Tufted Marbled Skipper

Genus *Spialia*
Spialia Sertorius Red-underwing Skipper

Genus *Pyrgus*
Pyrgus alveus Large Grizzled Skipper
Pyrgus andromedae Alpine Grizzled Skipper
Pyrgus armoricanus Orberthür's Grizzled Skipper
Pyrgus carthami Safflower Skipper
Pyrgus centaureae Northern Grizzled Skipper
Pyrgus malvae Grizzled Skipper
Pyrgus serratulae Olive Skipper

Genus *Carterocephalus*
Carterocephalus palaemon Chequered Skipper
Carterocephalus silvicolus Northern Chequered Skipper

Genus *Heteropterus*
Heteropterus morpheus Large Chequered Skipper

Genus *Thymelicus*
Thymelicus acteon Lulworth Skipper
Thymelicus lineola Essex Skipper
Thymelicus sylvestris Small Skipper

Genus *Hesperia*
Hesperia comma Silver-spotted Skipper

Genus *Ochlodes*
Ochlodes sylvanus Large Skipper

Family Papilionidae

Genus *Iphiclides*
Iphiclides podalirius Scarce Swallowtail

Genus *Papilio*
Papilio machaon Swallowtail

Genus *Parnassius*
Parnassius apollo Apollo
Parnassius mnemosyne Clouded Apollo

Family Pieridae

Genus *Leptidea*
Leptidea sinapis/Leptidea juvernica Wood White/Cryptic Wood White

Genus *Anthocharis*
Anthocharis cardamines Orange-tip

Genus *Aporia*
Aporia crataegi Black-veined White

Genus *Pieris*
Pieris brassicae Large White

Pieris napi Green-veined White
Pieris rapae Small White

Genus *Pontia*
Pontia daplidice/Pontia edusa Bath White/Eastern Bath White

Genus *Colias*
Colias crocea Clouded Yellow
Colias hecla Northern Clouded Yellow
Colias hyale/Colias alfacariensis Pale Clouded Yellow/Berger's Clouded Yellow
Colias myrmidone Danube Clouded Yellow
Colias palaeno Moorland Clouded Yellow
Colias tyche (previously *nastes*) Pale Arctic Clouded Yellow

Genus *Gonepteryx*
Gonepteryx rhamni Brimstone

Family Riodinidae

Genus *Hemearis*
Hemearis lucina Duke of Burgundy

Family Lycaenidae

Genus *Callophrys*
Callophrys rubi Green Hairstreak

Genus *Thecla*
Thecla betulae Brown Hairstreak

Genus *Favonius*
Favonius quercus Purple Hairstreak

Genus *Satyrium*
Satyrium ilicis Ilex Hairstreak
Sayrium pruni Black Hairstreak
Satyrium spini Blue-spot Hairstreak
Satyrium w-album White-letter Hairstreak

Genus *Lycaena*
Lycaena alciphron Purple-shot Copper
Lycaena dispar Large Copper
Lycaena helle Violet Copper
Lycaena hippothoe Purple-edged Copper
Lycaena phlaeas Small Copper
Lycaena tiyrus Sooty Copper
Lycaena virgaureae Scarce Copper

Genus *Lampides*
Lampides boeticus Long-tailed Blue

Genus *Cupido*
Cupido minimus Small Blue
Cupido argiades Short-tailed Blue

Genus *Celastrina*
Celastrina argiolus Holly Blue

Genus *Phengaris* (formerly *Maculinea*)
Phengaris alcon Alcon Blue

Phengaris arion Large Blue
Phengaris nausithous Dusky Large Blue
Phengaris telejus Scarce Large Blue

Genus *Glaucopsyche*
Glaucopsyche alexis Green-underside Blue

Genus *Pseudophylotes*
Pseudophylotes vicrama Eastern Baton Blue

Genus *Scolitantides*
Scolitantides orion Chequered Blue

Genus *Plebejus*
Plebejus aquilo Arctic Blue
Plebejus argus Silver-studded Blue
Plebejus argyrognomon Reverdin's Blue
Plebejus idas Idas Blue
Plebejus optilete Cranberry Blue
Plebejus orbitulus Alpine Blue

Genus *Aricia*
Aricia agestis Brown Argus
Aricia artaxerxes Northern Brown Argus
Aricia eumedon Geranium Argus
Aricia nicias Silvery Argus

Genus *Polyommatus*
Polyommatus amandus Amanda's Blue
Polyommatus bellargus Adonis Blue
Polyommatus coridon Chalkhill Blue
Polyommatus damon Damon Blue
Polyommatus dorylas Turquoise Blue
Polyommatus icarus Common Blue
Polyommatus semiargus Mazarine Blue

Family Nymphalidae, subfamily Nymphalinae

Genus *Limenitis*
Limenitis camilla White Admiral
Limenitis populi Poplar Admiral

Genus *Apatura*
Apatura iris Purple Emperor
Apatura ilia Lesser Purple Emperor

Genus *Araschnia*
Araschnia levana Map

Genus *Vanessa*
Vanessa atalanta Red Admiral
Vanessa cardui Painted Lady

Genus *Nymphalis*
Nymphalis antiopa Camberwell Beauty
Nymphalis polychloros Large Tortoiseshell
Nymphalis vau-album False Comma
Nymphalis xanthomelas Scarce Tortoiseshell

Genus *Polygonia*
Polygonia c-album Comma

Genus *Aglais*
Aglais io Peacock
Aglais urticae Small Tortoiseshell

Genus *Argynnis*
Argynnis adippe High Brown Fritillary
Argynnis aglaja Dark Green Fritillary
Argynnis laodice Pallas's Fritillary
Argynnis niobe Niobe Fritillary
Argynnis paphia Silver-washed Fritillary

Genus *Issoria*
Issoria lathonia Queen of Spain Fritillary

Genus *Brethnis*
Brethnis daphne Marbled Fritillary
Brethnis hecate Twin-spot Fritillary
Brethnis ino Lesser Marbled Fritillary

Genus *Euphydryas*
Euphydryas aurinia Marsh Fritillary

Euphydryas iduna Lapland Fritillary
Euphydryas maturna Scarce Fritillary

Genus *Boloria*
Boloria aquilonaris Cranberry Fritillary
Boloria chariclea Arctic Fritillary
Boloria dia Weaver's Fritillary
Boloria eunomia Bog Fritillary
Boloria euphrosyne Pearl-bordered Fritillary
Boloria freija Freija's Fritillary
Boloria frigga Frigga's fritillary
Boloria improba Dusky-winged Fritillary
Boloria napaea Mountain Fritillary
Boloria polaris Polar Fritillary
Boloria selene Small Pearl-bordered Fritillary
Boloria thore Thore's Fritillary
Boloria titania Titania's Fritillary

Genus *Melitaea*
Melitaea athalia Heath Fritillary
Melitaea aurelia Nickerl's Fritillary
Melitaea britomartis Assman's Fritillary
Melitaea cinxia Glanville Fritillary
Melitaea diamina False Heath Fritillary
Melitaea didyma Spotted Fritillary
Melitaea phoebe Knapweed Fritillary

Family Nymphalidae, subfamily Satyrinae

Genus *Pararge*
Pararge aegeria Speckled Wood

Genus *Lasiommata*
Lasiommata maera Large Wall Brown
Lasiommata megera Wall Brown
Lasiommata petropolitana Northern Wall Brown

Genus *Lopinga*
Lopinga achine Woodland Brown

Genus *Pyronia*
Pyronia tithonus Gatekeeper

Genus *Maniola*
Maniola jurtina Meadow Brown

Genus *Hyponephele*
Hyponephele lycaon Dusky Meadow Brown

Genus *Aphantopus*
Aphantopus hyperantus Ringlet

Genus *Coenonympha*
Coenonympha arcania Pearly Heath
Coenonympha glycerion Chestnut Heath
Coenonympha hero Scarce Heath
Coenonympha pamphilus Small Heath
Coenonympha tullia Large Heath

Genus *Erebia*
Erebia aethiops Scotch Argus
Erebia disa Arctic Ringlet
Erebia embla Lapland Ringlet
Erebia epiphron Mountain Ringlet
Erebia ligea Arran Brown
Erebia medusa Woodland Ringlet
Erebia pandrose Dewy Ringlet
Erebia polaris Arctic Woodland Ringlet

Genus *Melanargia*
Melanargia galathea Marbled White

Genus *Hipparchia*
Hipparchia alcyone Rock Grayling
Hipparchia semele Grayling
Hipparchia statilinus Tree Grayling

Genus *Oeneis*
Oeneis bore Arctic Grayling
Oeneis jutta Baltic Grayling
Oeneis norna Norse Grayling

Further Reading

NORTHERN EUROPEAN BUTTERFLIES

Dal, B. 1978, 1982 *The Butterflies of Northern Europe*. London & Canberra: Croom Helm
Fine early work, organized according to habitats, with paintings of butterflies as in life.

Eliasson, C. U. 2009 *Dägfjarilar I Orebro och Väsmanlands Iän*. Västerås: Västra Aros AB
Beautifully illustrated and informative book, covering parts of northern Europe, but without English-language text.

Eliasson, C. U., Ryrholm, N., Holmer, M., Jilg, K. & Gärdenfors, U. 2005 *Nationalnyckeln Till Sveriges Flora och Fauna. Fjärilar: Dagfjärilar*. Uppsala: Artdatabanken, SLU
Large-scale, fully comprehensive treatment of Scandinavian butterflies, with each species illustrated, keys, life histories and distribution maps. Mostly in Swedish, but with brief summaries in English.

Henriksen, H. J. & Kreutzer, I. B. 1982 *The Butterflies of Scandinavia in Nature*. Odense, Denmark: Skandinavisk Bogforlag
Inspirational, pioneering work, using photographic illustrations of living butterflies and their habitats, and breaking away from the practice of illustrating dead, set specimens.

Lafranchis, T. 2000 *Les Papillons du Jour de France, Belgique et Luxembourg et Leurs Chenilles*. Méze (France). Éditions Biotope
Beautifully illustrated work by outstanding field naturalist, with keys and distribution maps. Text in French.

Stoltze, M. 1996 *Danske Dagsommerfugle*. Copenhagen: Gyldendalske Boghandel, Nordisk Forlag A. S.
Finely illustrated and informative book, covering parts of northern Europe. No English-language text.

EUROPEAN BUTTERFLIES

There are numerous works, here is just a small selection.

Benton, T. & Bernhard, T. 2006 *Easy Butterfly Guide*. London: Aurum
Includes a selection of 123 European species, together with 'look-alikes', with extensive text for each species and distributions. Each species is illustrated by a large photograph, together with Tim Bernhard's superb paintings of the caterpillars and pupae.

Chinery, M. 1998 A *Photographic Guide to the Butterflies of Britain and Europe* (photos by T. Benton, B. Yates-Smith and others). London: HarperCollins
Comprehensive treatment by one of the UK's foremost natural history authors, with distribution maps and photographic illustrations of live butterflies.

Haahtela, T., Saarinen, K., Ojalainen, P. & Aarnio, H. 2011 *Butterflies of Britain and Europe: A Photographic Guide*. London: A. & C. Black
This fully comprehensive treatment of European butterflies is wider in geographical scope than most comparable books, and has superb colour photographs of all species, up-to-date distribution maps, identification clues and a very informative text. The Finnish authors are especially useful on northern species.

Kudrna, O., Harpke, A., Lux, K., Pennerstorfer, J., Schweiger, O., Settele, J. & Wiemers, M. 2011, 2015 *Distribution Atlas of Butterflies in Europe*. Halle, Germany: Gesellschaft für Schmetterlingschutz
Van Swaay, C. A. M. & Warren, M. S. 1999 *Red Data Book of European Butterflies (Rhopalocera)*. Nature and Environment, N. 99. Strasbourg: Council of Europe
Together, these compendious publications gather information about the distribution of European butterflies, and their conservation status in Europe as a whole, as well as in individual countries.

Lafranchis, T. 2004 *Butterflies of Europe*. Paris: Diatheo (French edition 2007)
Organized as an identification key, an authoritative field guide with photographic illustrations of live butterflies, summary information about each species and distribution maps.

Settele, J, Shreeve, T., Konvi ka, M. & Van Dyck, H. (eds) 2009 *Ecology of Butterflies in Europe*. Cambridge: Cambridge University
Wide-ranging and quite technical update of the scientific research about butterflies, for readers who would like to take their knowledge of the topic further.

Tolman, T. & Lewington, R. 1997 and later *Collins Field Guide: Butterflies of Britain and Europe*. London: HarperCollins
The outcome of many years' intensive fieldwork throughout Europe, full of information about life histories, behaviour and ecology, with distribution maps and superb paintings of each species by Lewington.

■ Further Reading ■

BUTTERFLIES IN BRITAIN AND IRELAND

There are many books on the whole of Britain and Ireland, and also numerous excellent books on particular counties and regions. My own favourites include:

Barkham, P. 2010 *The Butterfly Isles*. London: Granta
Combined butterfly guide and literary classic.

Dennis, R. L. H. 1977 *The British Butterflies: Their Origin and Establishment*. Oxford: E. W. Classey
Outstanding but very technical discussion of the history of the arrival and distribution of the British butterfly fauna.

Newland, D. E. 2006 *Discover Butterflies in Britain*. Hampshire: Wildguides
Essential guide to where to find the species you want to see.

Sandars, E. 1939 *A Butterfly Book for the Pocket*. London, etc.: Oxford University
My childhood inspiration.

Thomas, J. & Lewington, R. 2014 *The Butterflies of Britain and Ireland*. Oxford: British Wildlife
Can't be praised enough: it has everything.

WEBSITES

www.butterfly-conservation.org
This is the main UK-based organization for butterfly conservation, and hosts a group that is concerned with butterflies across mainland Europe. This is the European Interests Group (EIG), and its web address is www.bc-eig.org.uk. The EIG website carries a list of other websites relating to butterflies and moths for Europe generally, as well as for individual countries. These include two very valuable UK-based sources:

European Butterflies: A Portrait in Photographs (Bernard Watts- EIG)
http://butterflyeurope.co.uk

euroButterflies (Matt Rowlings – EIG)
www.eurobutterflies.com/index.php

ACKNOWLEDGEMENTS

In more than 35 years of exploring some of the more remote and often beautiful butterfly habits of Europe, I have encountered instances of kindness and hospitality just too numerous to mention, but I can mention some of the great companions who have shared these experiences and/or provided helpful advice: John Kramer, Simon Randolph, Vic Barnham, Tom Tolman, Andrew Wakeham-Dawson, Tim Bernhard, Matt Rowlings, Tristan Lafranchis, Claes Eliasson, Miguel Munguira, Sven Wair, Helmut Hottinger, the late Roy Cornhill and numerous others. Over recent decades most expeditions have been in the convivial company of Bernard Watts, from whose many-dimensional expertise I have benefited greatly. To my family, Shelley, Jay and Rowan, I owe both apologies and grateful thanks for tolerating the diversion of normal holiday expectations into the pursuit of butterflies, and for repeat absenteeism. Special mention goes to my incomparable life partner Shelley, who has, astonishingly, stayed the course for so long. I know it has not been easy! I am very grateful to the following for giving me permission to use their photographs in this book: Prof. Alan Dawson (*Nymphalis xanthomelas*) and Dr Bernard Watts (*Nymphalis vau album*, *Nymphalis antiopa* (upperside), *Cupido argiades* (male upperside), *Glaucopsyche alexis* (male upperside), *Aricia nicias* (male upperside).

■ INDEX ■